ENERGY METHODS IN CONTINUUM MECHANICS

Energy Methods in Continuum Mechanics

Proceedings of the Workshop on Energy Methods for Free Boundary Problems in Continuum Mechanics, held in Oviedo, Spain, March 21–23, 1994

Edited by

S. N. Antontsev

Lavrentyev Institute of Hydrodynamics,
Novosibirsk,
Russia

J. I. Díaz

Universidad Complutense de Madrid,
Spain

and

S. I. Shmarev

Lavrentyev Institute of Hydrodynamics,
Novosibirsk,
Russia

KLUWER ACADEMIC PUBLISHERS

DORDRECHT / BOSTON / LONDON

A C.I.P. Catalogue record for this book is available from the Library of Congress

ISBN 0-7923-4229-1

Published by Kluwer Academic Publishers,
P.O. Box 17, 3300 AA Dordrecht, The Netherlands.

Kluwer Academic Publishers incorporates
the publishing programmes of
D. Reidel, Martinus Nijhoff, Dr W. Junk and MTP Press.

Sold and distributed in the U.S.A. and Canada
by Kluwer Academic Publishers,
101 Philip Drive, Norwell, MA 02061, U.S.A.

In all other countries, sold and distributed
by Kluwer Academic Publishers Group,
P.O. Box 322, 3300 AH Dordrecht, The Netherlands.

Printed on acid-free paper

Printed in the Netherlands

Table of contents

Preface

This volume contains the proceedings of the Workshop *Energy Methods for Free Boundary Problems in Continuum Mechanics*, held in Oviedo, Spain, from March 21 to March 23, 1994.

It is well known that the conservation laws and the constitutive equations of Continuum Mechanics lead to complicated coupled systems of partial differential equations to which, as a rule, one fails to apply the techniques usually employed in the studies of scalar uncoupled equations such as, for instance, the maximum principle. The study of the qualitative behaviour of solutions of the systems requires different techniques, among others, the so called, *Energy Methods* where the properties of some integral of a nonnegative function of one or several unknowns allow one to arrive at important conclusions on the envolved unknowns. This volume presents the state of the art in such a technique. A special attention is paid to the class of Free Boundary Problems.

The organizers are pleased to thank the European Science Foundation (Program on Mathematical treatment of free boundary problems), the DGICYT (Spain), the FICYT (Principado de Asturias, Spain) and the Universities of Oviedo and Complutense de Madrid for their generous financial support. Finally, we wish to thank Kluwer Academic Publishers for the facilities received for the publication of these Proceedings.

S.N. Antontsev,
J.I. Díaz,
S.I. Shmarev.

QUASILINEAR PARABOLIC EQUATIONS WITH NON-ISOTROPIC NONLINEARITIES: SPACE AND TIME LOCALIZATION

S.N.ANTONTSEV
Lavrentyev Institute of Hydrodynamics
Lavrentyev Prospect 15, Novosibirsk, 630090, Russia
and University of Oviedo, c/Calvo Sotelo, s/n
Oviedo, 33007, Spain

Abstract.

The energy method is used to study some support properties of local solutions of second-order nonlinear parabolic equations with non-isotropic nonlinearities with respect to the solution and its spatial derivatives. We establish such properties of local weak solutions as finite speed of propagations of the initial disturbances, the waiting time phenomenon, and stable localization. The conditions providing these effects are formulated in terms of local assumptions on the data and the character of nonlinearity of the equation under consideration.

1. Introduction

1.1. STATEMENT OF THE PROBLEM. FORMULATION OF RESULTS.

In this paper we shall consider the propagation and vanishing properties of local weak solutions of nonlinear parabolic equations which admit non-isotropic degeneracy with respect to the solutions and its spatial derivatives. Let $\Omega \subset \mathbf{R}^N$, $N = 1, 2, \ldots$, be an open connected bounded domain with the boundary $\partial\Omega$, and $T \in (0, \infty)$. We consider the problem

$$\left.\begin{array}{c} \left(|u|^{\gamma-1}u\right)_t = \operatorname{div}\left(\vec{A}(x,t,u,\nabla u)\right) \\ -B(x,t,u) + f(x,t) \quad \text{in} \quad Q = \Omega \times (0,T), \\ u(x,0) = u_0(x) \quad \text{in} \quad \Omega, \end{array}\right\} \tag{1}$$

S. N. Antontsev et al. (eds.), *Energy Methods in Continuum Mechanics*, 1–12.

assuming that the functions $\vec{A}$ and B are subject to the following structural conditions:

$$\begin{aligned}
\forall(x,t,s,q) &\in \Omega\times\mathbf{R}^+\times\mathbf{R}\times\mathbf{R}^N \\
M_2\sum_{i=1}^{N}|q_i|^{p_i} &\le (\vec{A}(x,t,s,\rho),\rho)\le M_1\sum_{i=1}^{N}|q_i|^{p_i}, \qquad (2)\\
M_3|s|^{\sigma+1} &\le sB(x,t,s), \qquad (3)
\end{aligned}$$

with some positive constants $\gamma>0$, $\sigma>0$, M_i, $i=1,2,3$, and $p_i>1$, $i=1,...,N$.

A particular case of (1) is the equation

$$\frac{\partial}{\partial t}\left(u|u|^{\gamma-1}\right)=\sum_{i=1}^{N}\frac{\partial}{\partial x_i}\left(\left|\frac{\partial u}{\partial x_i}\right|^{p_i-2}\frac{\partial u}{\partial x_i}\right)-M_3\,u\,|u|^{\sigma-1}+f(x,t), \qquad (4)$$

$$1<p_i<\infty,\quad 0<(\gamma,\quad\sigma)<\infty.$$

The right-hand side $f(x,t)$ of equation (1) and the initial data $u_0(x)$ are assumed to satisfy

$$u_0\in L^{\alpha+1}(\Omega),\quad f\in L^{(1+\gamma)/\gamma}(Q)\quad\text{or}\quad f\in L^{(1+\sigma)/\sigma}(Q), \qquad (5)$$

We are interested in the qualitative properties of solutions of problem (1), understood in the following sense.

Definition 1 *A measurable in Q function $u(x,t)$, bounded along with its generalized derivatives $\partial u/\partial x_i$, $i=1,...,N$, is said to be a weak solution of problem (1) if:*

a) $\lim_{t\to 0}\|u(x,t)-u_0(x)\|_{L^{\gamma+1}(\Omega)}=0$;

b) *for any test function $\zeta(x,t)\in W^{1,\infty}(0,T:\overset{\circ}{W}{}^{1,p}(\Omega))$, vanishing at $t=T$, the integral identity holds*

$$\int_Q\left\{u\,|u|^{\gamma-1}\,\zeta_t-\vec{A}(x,\,t,\,u,\,\nabla\,u)\,\nabla\,\zeta+M_3\,u\,|u|^{\sigma-1}\,\zeta+f\,\zeta\right\}dxdt$$
$$+\int_\Omega u_0\zeta(x,0)dx=0. \qquad (6)$$

Let us note at once that we will never touch any question concerning the solvability of problem (1). As we said, the paper deals with qualitative properties of local weak solutions of problem (1), regardless the boundary conditions on $\Sigma = (0,T) \times \Omega$ they probably correspond to. Evidently, each of the functions satisfying in a weak sense equation (1), some boundary conditions on Σ, and the initial conditions is also a *local* weak solution in the sense of our Definition.

By now, the theory of problems of the type (1) already accounts for a number of the existence results. We refer the reader to papers [12, 13, 14, 15] and their references.

Let us accept the following notations

$$B_\rho(x_0) = \{x \in \Omega : \quad |x - x_0| < \rho\}, \quad \left((x_0, \quad \overline{B}_\rho(x_0)) \in \Omega\right).$$

Our objective is to prove the assertions.

Theorem 1 *(Finite Speed of Propagation-FSP. Waiting Time Phenomen-WTP.) Let $u(x,t) \not\equiv 0$ be a weak solution to problem (1), with the data $f(x,t) \equiv 0$ and*

$$u(x,0) = 0 \quad x \in B_{\rho_0}(x_0), \quad (\overline{B}_{\rho_0}(x_0) \in \Omega) \tag{7}$$

such that

$$(|u(x,t)|, \quad |\nabla_x u(x,t)|) \leq M, \tag{8}$$

Assume that the structural constants γ, p_i, $i = 1, .., N$ satisfy the conditions

$$(\beta - \alpha)\frac{\theta(1+\gamma)}{\beta(1-\theta)} + 1 + \gamma < \alpha, \quad 0 \leq M_3, \tag{9}$$

with $\alpha = \min_i p_i, \beta = \max_i p_i$ and

$$\theta = \left(\frac{1}{1+\gamma} - \frac{N-1}{N\beta}\right) / \left(\frac{1}{N} - \frac{p^*}{N} + \frac{1}{1+\gamma}\right), \tag{10}$$

$$0 < \theta < 1, \quad p^* = \sum_i^N \frac{1}{p_i}$$

Then $u(x,t)$ possesses the FSP property, i.e there exist an instant $t^* = t^*(M, \rho_0, \gamma, p_i, T) > 0$ and a function $\rho(t)$, $(\rho(t) \leq \rho_0$, $\rho(0) = \rho_0$, $0 \leq t \leq t^* \leq T)$ such that

$$u(x,t) = 0, \quad |x| \leq \rho(t), \quad 0 \leq t \leq t^*$$

If $f(x,t) \not\equiv 0$ and , in addition to (7), the data of the problem satisfy the condition

$$\left(\int_{B_\rho}|u_0(x)|^{1+\gamma}dx \quad + \quad \int_0^T \|f(\cdot,t)\|^{(1+\gamma)/\gamma}_{L^{(1+\gamma)/\gamma}(B_\rho)}dt\right)$$

$$\leq \quad \varepsilon(\rho_0-\rho)_+^{\frac{1}{1-\nu}}, \quad \rho_0 \leq \rho \tag{11}$$

$$\nu = \left(1-\frac{1-\theta}{1+\gamma}-\frac{\theta}{\beta}\right)\frac{\alpha}{\alpha-1}, \tag{12}$$

then $u(x,t)$ possesses the WTF property, e.g., there exists $t^* > 0$ such that

$$u(x,t)=0, \quad |x|\leq\rho_0, \quad 0\leq t\leq t^*.$$

Theorem 2 *(Stable Localization-SL.) Let $u(x,t)\not\equiv 0$ be a weak solution to problem* (1)*, with the data $f(x,t)\not\equiv 0$ and*

$$u(x,0)=0 \quad x\in B_{\rho_0}(x_0), \quad (\overline{B}_{\rho_0}(x_0)\in\Omega) \tag{13}$$

such that

$$(|u(x,t)|, \quad |\nabla_x u(x,t)|)\leq M, \tag{14}$$

Assume that the structural constants satisfy the conditions

$$1+\sigma<\beta, \quad \sigma<\gamma, \quad 0<M_3, \quad (\beta-\alpha)(1+\gamma)/\alpha<(1-\theta)(\beta-1-\sigma), \tag{15}$$

with $\alpha=\min_i p_i, \beta=\max_i p_i$ and

$$\theta=\left(\frac{1}{\mu}-\frac{N-1}{N\beta}\right)/\left(\frac{1}{N}-\frac{p^*}{N}+\frac{1}{\mu}\right), \tag{16}$$

$$0<\theta<1, \quad p^*=\sum_i^N\frac{1}{p_i}, \quad \mu=\frac{\beta(1+\sigma)}{\beta+\gamma-\sigma}$$

Assume also that the data of the problem satisfy the condition

$$\left(\int_{B_\rho}|u_0(x)|^{1+\gamma}dx \quad + \quad \int_0^T \|f(\cdot,t)\|^{(1+\sigma)/\sigma}_{L^{(1+\sigma)/\sigma}(B_\rho)}dt\right)$$

$$\leq \quad \varepsilon(\rho_0-\rho)_+^{\frac{1}{1-\nu}}, \quad \rho_0\leq\rho \tag{17}$$

with

$$\nu = \left(\frac{\beta - 1}{\beta} - \frac{(1-\theta)(\theta - 1 - \sigma)}{\beta(1+\gamma)}\right)\frac{\alpha}{\alpha - 1}, \tag{18}$$

Then there exist positive constants M_*, ε_* such that , if $M \leq M_*$, $\varepsilon \leq \varepsilon_*$, then $u(x,t)$ possesses the SL property, i.e

$$u(x,t) = 0, \quad |x| \leq \rho_0, \quad 0 \leq t \leq T$$

Remark 1 *We remark that if $\alpha = \beta = p_i = p$, $i = 1, ..., N$, then conditions* (9) *coincide with the already known condition $p > 1 + \gamma$ that is valid in the case of non-isotrophic degeneracy (see [12]). It is easy to see that the first of conditions* (9) *and the last of* (15) *are surely fulfilled if the difference $\beta - \alpha$ is small enough.*

Let us also mention that the first of the assertion of Theorem 1 (the FSP property) was preliminary formulated without proof in [18].

Remark 2 *Condition* (11) *poses the constraint of appropriate rate of vanishing as $\rho \to \rho_0 + 0$ (in the integrale sense) of the initial function $u_0(x)$ and the source function $f(x,t)$.*

Let us show that condition (11), *(with $f \equiv 0$), is sharp in case $N = 1$. Let $v(x,t)$ be a bounded nonnegative solution of the problem ($k > 1$, $m > 1$)*

$$\frac{\partial v}{\partial t} = \frac{\partial}{\partial x}\left(\left|\frac{\partial(v|v|^{m-1})}{\partial x}\right|^{k-1}\frac{\partial(v|v|^{m-1})}{\partial x}\right), \tag{19}$$

$$v(x,0) = v_0(x) \geq 0 \quad x = x_1 \in \Omega = (-L, L), \tag{20}$$

$$(v_0 = 0, \ |x| \leq \rho_0 < L)$$

It is known, [12], *that the condition*

$$v_0(x) \leq C \leq |x - \rho_0|^{\kappa} \qquad \left(C = const., \kappa = \frac{k+1}{km - 1}\right) \tag{21}$$

is necessary and sufficient (with respect to the order κ) for $v(x,t)$ to display the metastable localization effect. (In terms of [12], *to provide inertion of propagation of disturbances across the point $x = \rho_0$).*

Having been recalculated for the solution $u(x,t)$ of problem (1), *condition* (21) *takes the form*

$$|u_0(x)| \leq C \leq |x - \rho_0|^{\kappa/\gamma}, \quad \kappa/\gamma = \frac{p}{p - \gamma - 1}, \quad p = p_1.$$

It is easy to verify that the last condition proves necessary for (11), *since*

$$\int_{\rho_0}^{\rho} |u_0(x)|^{1+\gamma}dx \le C^{1+\gamma}(1-\mu)(\rho-\rho_0)^{1/(1-\nu)}, \quad \rho \in (\rho_0, L).$$

It is to be noted that (11) , *analogously to* ([6, 8, 9]), *may be replaced by the assumption of convergence of the integral*

$$I: = \int_{\rho_0+0} (\rho-\rho_0)^{\lambda} \Big(\|u_0\|^{1+\gamma}_{L^{1+\gamma}(B_\rho)} + \|f\|^{(1+\gamma)/\gamma}_{L^{(1+\gamma)/\gamma}(B_\rho \times (0,T))} \Big)^{p/(p-1)} d\rho < \infty$$

with some constant $\lambda = \lambda(\gamma, p)$

2. Proofs of Theorems 1,2.

2.1. DERIVATION AND ANALYSIS OF DIFFERENTIAL INEQUALITIES.

To prove Theorems 1, 2 we will use the energy method which was proposed in [1] and developed in ([4]-[9],[11, 16, 17]). We introduce the energy functions

$$b(\rho,t) = \int_{B_\rho} |u(x,t)|^{1+\gamma}dx, \quad \bar{b}(\rho,t) = ess \sup_{0 \le \tau \le t} b(\rho,\tau) \tag{22}$$

$$E(\rho,t) = \int_0^t \int_{B_\rho} \vec{A}\nabla u dx dt \equiv \sum_{i=1}^{N} E_i, \quad C(\rho,t) = \int_0^t \int_{B_\rho} |u|^{1+\sigma} dx dt \tag{23}$$

For the sake of simplicity and without any loss of generality, we will assume further that $x_0 = 0$, $B_\rho(0) = B_\rho$.

It is easy to check the following properties of function $E(\rho,t)$:

$$M_2 \sum_{i=N}^{N} \|\partial u/\partial x_i\|^{p_i}_{L^{p_i}(B_\rho)} \le E(\rho,t) \le M_1 \sum_{i=N}^{N} \|\partial u/\partial x_i\|^{p_i}_{L^{p_i}(B_\rho)} \tag{24}$$

$$\frac{\partial E(\rho,t)}{\partial \rho} \equiv E'_\rho = \int_0^t \int_{S_\rho} \vec{A}\nabla u dx dt \equiv \sum_{i=1}^{N} E'_{i\rho}, \quad S_\rho = \partial B_\rho \tag{25}$$

$$\max_{0\le\tau\le t}\frac{\partial E(\rho,\tau)}{\partial\rho}=\frac{\partial(\max_{0\le\tau\le t}E(\rho,\tau))}{\partial\rho},\quad \frac{\partial E(\rho,\tau)}{\partial t}\equiv E'_t=\int_{B_\rho}\vec{A}\nabla u\,dx\,dt.$$

Step 1. Using the standard way (see, for example [2, 11]) it is easy to infer from (6) that, in terms of these functions, every weak solution $u(x,t)$ of problem (1) fulfils the energy relation

$$\frac{\gamma}{1+\gamma}b(\rho,t)+E(\rho,t)+M_3C(\rho,t)\le J_1+J_2, \tag{26}$$

where

$$J_1=\int_0^t\int_{S_\rho}u\sum_{i=1}^N\left|\frac{\partial u}{\partial x_i}\right|^{p_i-2}\frac{\partial u}{\partial x_i}\nu_i\,dS\,d\tau,$$

$$J_2=\frac{\gamma}{1+\gamma}b(\rho,0)+\int_0^t\int_{B_\rho}uf\,dx\,d\tau,$$

and $\vec{\nu}=(\nu_1,...,\nu_N)$ is the outward normal vector to $S_\rho=\partial B_\rho$.

Further, let us make use of the following assertions

Lemma 1 ([18]) *Let $\Omega\subset R^N$ be a bounded domain with the smooth boundary $\Gamma=\partial\Omega$,*

$$u(x)\in W^{1,,\infty}(\Omega),\quad |\nabla u|\le M.$$

Then

$$\|u\|_{L^q(\Gamma)}\le K\left(\sum_{i=1}^N\|\frac{\partial u}{\partial x_i}\|_{L^{p_i}(\Omega)}+\|u\|_{L^r(\Omega)}\right)^\theta\|u\|_{L^r(\Omega)}^{(1-\theta)}, \tag{27}$$

where

$$K_0=K_0(M,\Omega,p_i,q,r),\quad q\ge p^*/N,\quad p^*=\sum_{i=1}^N 1/p_i$$

$$\theta=(1/q-(N-1)/N\beta)/(1/N-p^*/N+1/r),\quad q\ge r>1,\quad 0<\theta<1$$

Corollary 1 *If in Lemma 1 we suppose that $\Omega=B_\rho$, $S_\rho=\partial B_\rho$, $\rho>0$, then inequality (27) takes the form*

$$\|u\|_{L^q(S_\rho)}\le K_0\left(\sum_{i=1}^N\rho^{\delta_i}\|\frac{\partial u}{\partial x_i}\|_{L^{p_i}(B_\rho)}+\rho^{-\delta}\|u\|_{L^r(B_\rho)}\right)^\theta\|u\|_{L^r(B_\rho)}^{(1-\theta)} \tag{28}$$

where

$$\delta = \frac{N(q-r)+qr}{qr)}, \quad \delta_i = N\left(\frac{1}{q} - \frac{1}{p_i}\right)$$

and parameter θ is defined in Lemma 1.

Proof. We perform the following change of variable

$$x = \rho y, \quad x \in B_\rho(0), \quad y \in B_0$$

If $u(x) \in W^{1,\infty}(B_\rho(0))$, then the function $v(y) = u(x) \in W^{1,\infty}(B_1(0))$ and we get (28) using in (16) . the formulas

$$\|v\|_{L^q(B_1(0))} = \rho^{-N/q}\|u\|_{L^q(B_\rho(0))}, \quad \|v\|_{L^q(S_1(0))} = \rho^{-(N-1)/q}\|u\|_{L^q(S_\rho(0))},$$

$$\|\frac{\partial v}{\partial x_i}\|_{L^{p_i}(B_1(0))} = \rho^{1-N/p_i}\|\frac{\partial u}{\partial y_i}\|_{L^{p_i}(B_\rho(0))}.$$

We begin with the proof of the first of assertions of Theorem 1 (FSP property). Then in (26) $\rho \leq \rho_0$, $b(\rho, 0) = 0$, $f = 0$ and, correspondingly, $J_2 \equiv 0$. Using Hölder inequality and (8) gives the following estimate for J_1:

$$\begin{aligned} |J_1| &\leq \sum_{i=1}^{N}\left(\int_0^t\int_{S_\rho}|u|^{p_i}dSd\tau\right)^{1/p_i}\left(\int_0^t\int_{S_\rho}|u_{x_i}|^{p_i}dSd\tau\right)^{(p_i-1)/p_i} \\ &\leq \left(\int_0^t\int_{S_\rho}|u|^{\beta}dSd\tau\right)^{1/\beta}\sum_{i=1}^{N}(\text{meas } S_\rho)^{(\beta-p_i)/\beta}\left(\frac{\partial E_i}{\partial\rho}\right)^{(p_i-1)/p_i} \qquad (29) \\ &\leq K\left(\int_0^t\|u\|^{\beta}_{L^\beta(S_\rho)}d\tau\right)^{1/\beta}\left(\frac{\partial E}{\partial\rho}\right)^{(\alpha-1)/\alpha}. \end{aligned}$$

From now on, we denote by K_0 the constants dependent only on structural constants and Ω, ρ_0, M, by K_0 and use the notation K if they depend also on T.

Rising both parts (28) (with $q = \beta$ and $r = 1+\gamma$) to the power β, integrating in t and using Hölder's inequality and (8), we come to the following chain of inequalities

$$\begin{aligned} \left(\int_0^t\|u\|^{\beta}_{L^\beta(S_\rho)}d\tau\right)^{1/\beta} &\leq K\rho^{-\delta\theta}\left(\int_0^t\left(E_t + b^{\beta/(1+\gamma)}\right)^{\theta}b^{(1-\theta)\beta/(1+\gamma)}\right)^{1/\beta} \\ &\leq K\rho^{-\delta\theta}\overline{b}^{(1-\theta)/(1+\gamma)}t^{(1-\theta)(\beta}\left(E+\overline{b}\right)^{\theta/\beta}. \qquad (30) \end{aligned}$$

To derive it, we made use of the inequality $\beta \geq 1+\gamma$. Gathering (29), (30) we arrive at the final estimate for J_1

$$|J_1| \leq K\rho^{-\delta\theta} t^{(1-\theta)/\beta} \left(\frac{\partial E}{\partial \rho}\right)^{(\alpha-1)/\alpha} \bar{b}^{(1-\theta)/(1+\gamma)} \left(E+\bar{b}\right)^{\theta/\beta}. \tag{31}$$

Returning to (26) we have, correspondingly,

$$\begin{aligned} &\frac{\gamma}{1+\gamma} b(\rho,t) + E(\rho,t) + M_3 C(\rho,t) \\ &\leq K\rho^{-\delta\theta} t^{(1-\theta)/\beta} \left(\frac{\partial E}{\partial \rho}\right)^{(\alpha-1)/\alpha} \bar{b}^{(1-\theta)/(1+\gamma)} \left(E+\bar{b}\right)^{\theta/\beta} \end{aligned} \tag{32}$$

Since the right-hand side of the last inequality is nondecreasing with respect to t, it remains true if we replace $b(\rho,t)$ by $\bar{b}(\rho,t)$. Rising both parts (32) to the power $\alpha/(\alpha-1)$ we come to the inequality

$$E^{\nu} \leq \left(\bar{b}+E\right)^{\nu} \leq K t^{\kappa} \rho^{-\lambda} \frac{\partial E}{\partial \rho}, \tag{33}$$

where

$$\nu = \left(1 - \frac{1-\theta}{1+\gamma} - \frac{\theta}{\beta}\right) \frac{\alpha}{\alpha-1}, \quad \kappa = \alpha(1-\theta)/\beta(\alpha-1), \quad \lambda = \delta\theta\alpha/(\alpha-1)$$

Inequality (33) represents an ordinary differential inequality in the variable ρ for the energy function $E(\rho,t)$. The variable t (time) enters this inequality like a parameter.

It is easy to verify that $0 < \nu < 1$ if only condition (9) was fulfilled.

To prove the second assertion of the Theorem 1, (e.g., $J_2 \not\equiv 0$ in (26)), we use (11) to rely on the habitual estimate

$$\begin{aligned} |J_2| &\leq \frac{\gamma}{1+\gamma} \varepsilon \left(\rho-\rho_0\right)_+^{1/1-\nu} + t\bar{b} + K\int_0^t \|f(\cdot,\tau)\|^{(1+\gamma)/\gamma}_{L^{(1+\gamma)/\gamma}}(B_\rho) d\tau \\ &\leq K\varepsilon\left(\rho-\rho_0\right)_+^{1/1-\nu} + t\bar{b} \end{aligned} \tag{34}$$

Joining (26), (31), (34) with $2t < \gamma/(1+\gamma)$ leads to the nonhomogeneous ordinary differential inequality

$$\begin{aligned} E^{\nu} \leq \left(\bar{b}+E\right)^{\nu} &\leq K t^{\kappa} \rho^{-\lambda} \frac{\partial E}{\partial \rho} \\ &+ K\varepsilon^{\nu} \left(\rho-\rho_0\right)_+^{\nu/(1-\nu)} \end{aligned} \tag{35}$$

Now we pass to the proof of Theorem 2. We use (26) (with $M_3 > 0$, $\rho_0 \le \rho$, $J_2 \neq 0$) again and (29). But to estimate $\|u\|_{L^\beta(S_\rho)}$ in (29) we apply (28) (with $q = \beta$ and $r = \mu = \beta(1+\sigma)/(\beta+\gamma-\sigma)$) and the inequality

$$\|u\|_{L^\mu(B_\rho)} \le \|u\|^{(\beta-1-\sigma)/\beta}_{L^{1+\gamma}(B_\rho)} \|u\|^{(1+\sigma)/\beta}_{L^{1+\sigma}(B_\rho)} = b^{(\beta-1-\sigma)/\beta(1+\gamma)}\, C_t^{1/\beta}. \tag{36}$$

Then (28) may be rewritten in the following form

$$\begin{aligned} \|u\|^\beta_{L^\beta(S_\rho)} &\le K_0 \left(E_t + b^{(\beta-1-\sigma)/(1+\gamma)}\, C_t\right)^\theta b^{(1-\theta)(\beta-1-\sigma)/\beta(1+\gamma)}\, C_t^{1-\theta} \\ &\le K_0 (E_t + C_t)\, b^{(1-\theta)(\beta-1-\sigma)/(1+\gamma)} \end{aligned} \tag{37}$$

To derive it, we made use of (14), (15). Substituting (37) into (29) we come to

$$\begin{aligned} |J_1| &\le K_0 \left(\frac{\partial E}{\partial \rho}\right)^{(\alpha-1)/\alpha} (E+C)^{1/\beta}\, \bar{b}^{(1-\theta)(\beta-1-\sigma)/\beta(1+\gamma)} \\ &\le \frac{1}{4}\left(\frac{\gamma}{1+\gamma}\bar{b} + E + M_3 C\right) + K_0 \left(\frac{\partial E}{\partial \rho}\right)^{1/\nu} \end{aligned} \tag{38}$$

We estimate the term J_2 in (26) in the following way

$$\begin{aligned} |J_2| &\le \frac{\gamma}{1+\gamma}\varepsilon\,(\rho-\rho_0)_+^{1/1-\nu} + \frac{M_3}{4} C(\rho,t) + K_0 \int_0^t \|f(\cdot,\tau)\|^{(1+\sigma)/\sigma}_{L^{(1+\sigma)/\sigma}(B_\rho)} d\tau \\ &\le K_0 \varepsilon\,(\rho-\rho_0)_+^{1/1-\nu} + \frac{M_3}{4} C(\rho,t). \end{aligned} \tag{39}$$

Gathering (26), (38) and (39) we obtain the final the nonhomogeneous ordinary differential inequality (c.f with (35))

$$\begin{aligned} E^\nu \le \left(\bar{b} + E\right)^\nu &\le K_0 \frac{\partial E}{\partial \rho} \\ &+ K_0 \varepsilon^\nu (\rho-\rho_0)_+^{\nu/(1-\nu)} \end{aligned} \tag{40}$$

Recall that the constants in (38)-(40) do not depend on T.

The analysis of inequalities (33), (35), (40) is the last step of the proofs of Theorem 1, 2. The complete analysis of such inequalities was given in [2, 3, 6, 10, 11, 16]. We present here its sketch. We start with the simpler case –inequality (33). Integrating (33) in $\rho \in (\rho, \rho_0)$ we obtain the estimate

$$E^{1-\nu}(\rho,t) \le E^{1-\nu}(\rho_0,t) - Kt^{\kappa}\left(\rho_0^{\lambda+1} - \rho^{\lambda+1}\right) \tag{41}$$

and , correspondingly,

$$E(\rho(t),t) = 0, \quad u(x,t) \equiv 0, \quad |x| \le \rho(t)$$

if

$$\rho^{\lambda+1}(t) = \rho_0^{\lambda+1} - Kt^{\kappa}E^{1-\nu}(\rho_0,t).$$

With all evidence, $0 < \rho(t)$ if

$$0 < t < t^* = \left(\rho^{1+\lambda}E_0^{\nu-1}/K\right)^{1/\kappa}, \quad \left(E(\rho_0,t) \le E_0 = T\mathrm{meas}B_{\rho_0}\sum_{i=1}^{N} M^{p_i}\right).$$

Let us study inequalities (35), (40). It is well known that their solutions admit the estimate of the following type (see [3, 10]):

$$E(\rho,t) \le \Lambda\,(\rho-\rho_0)_+^{1/(1-\nu)}, \quad \left(\Lambda = cst., \ \rho \in (\rho_0,\rho_1), \ \overline{B}_{\rho_1} \in \Omega\right),$$

if t^* is small enough in case (35) and , respectively, if $t \in (0,T)$ is arbitrary and E_*, ε_*, are small enough in case (40), (recall that $E(\rho,t) \le E_0(M) \le E_*(M_*)$, $\varepsilon \le \varepsilon_*$).).

This completes the proofs of Theorems 1,2.

References

1. S.N.Antontsev On the localization of solutions of nonlinear degenerate elliptic and parabolic equations. *Dokl. Akad. Nauk. SSSR*, 1981, 260, n.6, pp.1289-1293. (in Russian; English translation in *Soviet Math. Dokl.*, 1981, 24, n.2, pp.420-424).
2. S.N.Antontsev Localization of solutions of degenerate equations of continuum mechanics. *Akad. Nauk. SSSR, Sibirsk. Otdel. Inst. Hydrodynamics, 1986, 109 pp. (Russian)*
3. S. N. Antontsev and J. I. Diaz New results on localization of solutions of nonolinear elliptic and parabolic equations obtained by the energy method. *Dokl. Akad.N auk SSSR. 1988, Vol. 303, No.3, pp. 320-325.(English translation in Soviet Math. Dokl. 1988, Vol.33, No.3,pp.535-539).*
4. S.N.Antontsev, J.I.Diaz and S.I.Shmarev New applications of energy methods to parabolic free boundary Problems. *Uspekhi Mat. Nauk*, 1991, 46, issue 6 (282), pp. 181-182. (in Russian; English translation: in *Russian Mathematical Surveys*, 1991, 46, n.6, pp.193-194).
5. S.N.Antontsev, J.I.Diaz and S.I.Shmarev New results of the character of localization of solutions for elliptic and parabolic equations. *Birkhauser-Verlag*, International Series of Numerical Mathematics, 1992, 106, pp. 59-66.
6. S.N.Antontsev, J.I.Diaz and S.I.Shmarev The support shrinking properties for solutions of quasilinear parabolic equations with strong absorption terms. (To appear).
7. S.N.Antontsev, S.I. Shmarev Localization of solutions of nonlinear parabolic equations with linear sources of general type. *Dinamica Sploshnoi Sredy*, Novosibirsk, 1989, 89, pp. 28-42.

8. S.N.Antontsev, S.I.Shmarev Local energy method and vanishing of weak solutions of nonlinear parabolic equations. *Dokl. Akad. Nauk SSSR*, 1991, 318, n.4. pp. 777-781. (in Russian; English translation: in *Soviet Math. Doklady*, 1991, 43).
9. S.N.Antontsev, S.I.Shmarev Local energy method and vanishing properties of weak solutions of quasilinear parabolic equations. In book: *Free boundary problems involving solids*. Proceedings of the International Colloquim " Free Boundary Problems: Theory and Applications". J.Chadam and H.Rasmussen eds., Pitman Res. Notes in Math. ser., 281, Longman. 1993, pp.2-5.
10. J.I. Diaz Nonlinear partial differential equations and free boundaries. *Vol.1. Elliptic Equations. Research Notes in Math. No. 106. Pitman. 1985.*
11. J.I.Diaz, L.Veron Local vanishing properties of solutions of elliptic and parabolic quasilinear equations. *Trans. Amer. Math. Soc.*, 1985, 290, n.2, pp. 787-814.
12. A.S.Kalashnikov Some questions of the qualitative theory of nonlinear degenerate second-order parabolic equations. *Russian Math Surveys*, 1987, V.42, n.2, pp.169-222.
13. O.A. Ladyzhenskaya, V.A. Solonnikov and N.N. Ural'tseva Linear and quasilinear equations of parabolic type. Trans. Math. Monographs. 23. Amer. Math. Soc. Providence.1968.
14. J.L Lions Some methods of solutions of nonlinear boundary value problems. Moscow. Mir. 1972. p.288.
15. A.A.Samarsky, V.A.Galaktionov, S.P.Kurdumov, A.M.Mikhaylov *Blow-up in Problems for Quasilinear Parabolic Problems*, Nauka, Moscow, 1987 (in Russian; English translation: Walter de Groyter, Berlin, to appear).
16. S.I.Shmarev Local properties of solutions of nonlinear equation of heat conduction with absoption. *Dinamika Sploshnoi Sredy*, Novosibirsk, 1990, 95, pp. 28-42. (in Russian)
17. S.I.Shmarev The effect of support shrinking for solutions of a degenerate parabolic equation. *Dinamica sploshnoi Sredy*, 1990, 97, pp.194-200. (in Russian)
18. Yu. G. Rykov On the finite velocity of propagation of perturbation and inertia for generalized soluitions of degenerating quasilinear equations. *Differential Equations and Their Applications*, Moscow, Gos. Univ.Moscow, 1984, pp.160-164.

ON THE BOUNDARY LAYER FOR DILATANT FLUIDS

S. N. ANTONTSEV
Lavrentyev Institute of Hydrodynamics
Lavrentyev Prospect 15, Novosibirsk, 630090, Russia
and Universidade da Beira Interior,
R. Marqués d'Avila e Bolama
6200 Covilhã, Portugal

J. I. DIAZ
Departamento de Matemática Aplicada,
Universidad Complutense de Madrid,
28040 Madrid, Spain

S. I. SHMAREV
Lavrentyev Prospect 15, Novosibirsk, 630090, Russia
and University of Oviedo, c/Calvo Sotelo, s/n
33007, Oviedo, Spain

1 Introduction: the boundary layer and the von Mises transformation.

This paper deals with the boundary layer associated to a class of non Newtonian fluids, i. e., fluids for which the stress tensor $\mathbf{T}$, at given temperature and pressure, is not a linear function of the spatial variation of the velocity $\mathbf{L} \doteq \nabla \mathbf{v}$. This class of fluids is relevant in many contexts: chemical engineering (polymer melts, polymer solutions, suspensions, lubricants, paints, etc.), liquid crystals, oriented media, fibrous media, animal blood etc. (see, e. g., Schowalter [28] and Narasimhan [17]). The above notion of non-Newtonian fluids fails to bound the subject. An important subclass is the so called *purely viscous non Newtonian fluids.* To introduce this notion we start from the *Reiner-Rivlin principle of material objectivity*

$$\mathbf{T} = -P\mathbf{I} + \phi_1(\mathrm{I}_1, \mathrm{I}_2)\mathbf{D} + \phi_2(\mathrm{I}_2, \mathrm{I}_3)\mathbf{D}^2,$$

where P is the pressure, $\mathbf{I}$ is the identity tensor and I_i $(i = 1, 2, 3)$ are the principal invariants of $\mathbf{D} \doteq \frac{1}{2}(\nabla \mathbf{v} + \nabla \mathbf{v}^T)$, the symmetric part of $\mathbf{L}$. The special case of ϕ_1 identically constant and $\phi_2 = 0$ corresponds to the case of incompressible Newtonian fluids. The more general case of $\phi_2 = 0$ and non-constant ϕ_1 defines

S. N. Antontsev et al. (eds.), Energy Methods in Continuum Mechanics, 13–21.

the class of purely viscous non Newtonian fluids (also called generalized Newtonian fluids). It is useful to introduce the *shear stress* function

$$\tau(\kappa) = \frac{1}{2}\phi_1(\kappa)\kappa$$

where κ represents the *shear rate*. The *Power-law* or *Ostwald-de Waele model* is the one associated to the case of

$$\tau(\kappa) = K\,|\kappa|^{p-2}\,\kappa.$$

where $p > 1$ is given as a constitutive property of the fluid. If $p = 2$ we find again the class of Newtonian fluids. The case of $p > 2$ corresponds to the so called *dilatant fluids* and the case $1 < p < 2$ to the *pseudoplastic fluids.*

The Navier-Stokes system associated to a two-dimensional stationary flow of a incompressible dilatant fluid is

$$u\frac{\partial u}{\partial x} + v\frac{\partial u}{\partial y} = -\frac{1}{\rho}\frac{\partial P}{\partial x} + \nu\frac{\partial}{\partial x}(|\mathbf{D}|^{p-2}\frac{\partial u}{\partial x}) + \frac{\nu}{2}\frac{\partial}{\partial y}(|\mathbf{D}|^{p-2}(\frac{\partial u}{\partial y} + \frac{\partial v}{\partial x}))$$

$$u\frac{\partial v}{\partial x} + v\frac{\partial v}{\partial y} = -\frac{1}{\rho}\frac{\partial P}{\partial y} + \frac{\nu}{2}\frac{\partial}{\partial x}(|\mathbf{D}|^{p-2}(\frac{\partial u}{\partial y} + \frac{\partial v}{\partial x})) + \nu\frac{\partial}{\partial y}(|\mathbf{D}|^{p-2}\frac{\partial v}{\partial y})$$

$$\frac{\partial u}{\partial x} + \frac{\partial v}{\partial y} = 0$$

where $\mathbf{v} = (u,v)$ is the velocity, P the pressure,

$$\mathbf{D} = \begin{pmatrix} \frac{\partial u}{\partial x} & \frac{1}{2}\left(\frac{\partial u}{\partial y} + \frac{\partial v}{\partial x}\right) \\ \frac{1}{2}\left(\frac{\partial u}{\partial y} + \frac{\partial v}{\partial x}\right) & \frac{\partial v}{\partial y} \end{pmatrix}$$

and

$$|\mathbf{D}|^2 = u_x^2 + \frac{1}{2}(u_y + v_x)^2 + v_y^2.$$

In 1904, L. Prandtl [22] studied the influence of viscosity in a Newtonian flow at high Reynolds number in the presence of an obstacle. If we assume that the flow is exterior to a body (here represented by the interval (0,X) in the x-axes) and that a representative value of the modulus of the velocity is V, then the Reynolds number is $R=\frac{VX}{\nu}$ (we can assume, for simplicity, that $\rho \equiv 1$). The transition from zero velocity at the wall to the free stream velocity (velocity of the outer flow) (U(x),0) takes place in a very thin layer: *the boundary layer.* To study such a layer, Prandtl used some simplifications. For instance, it is natural to expect that

$$\frac{\delta}{X} << 1,$$

where δ is the boundary layer thickness. It is not difficult to see that this property is equivalent to the condition

$$\left|\frac{\partial u}{\partial y}\right| >> \left|\frac{\partial u}{\partial x}\right|.$$

Using dimensional analysis it can be shown that under this condition

$$\left|\frac{\partial P}{\partial y}\right| << 1.$$

So, following Prandtl, we can assume the Bernouilli equation for the outer flow to be

$$U(x)\frac{dU}{dx}(x) = -\frac{dP}{dx}(x).$$

Neglecting smaller terms, the Navier-Stokes system leads to the following problem:

$$(PS)\begin{cases} u\frac{\partial u}{\partial x} + v\frac{\partial u}{\partial y} = UU_x + \nu\frac{\partial}{\partial y}\left(\left|\frac{\partial u}{\partial y}\right|^{p-2}\frac{\partial u}{\partial y}\right) & \text{in Q}, \\ \frac{\partial u}{\partial x} + \frac{\partial v}{\partial y} = 0 & \text{in Q}, \\ u(0,y) = u_0(y) & y > 0, \\ u(x,0) = 0, v(x,0) = v_0(x) & x \in (0,X), \\ u(x,y) \to U(x) \text{ as } y \to \infty & x \in (0,X). \end{cases}$$

where Q $\doteq \{(x,y) : 0 < x < X, 0 < y\}$. In most physical problems $v_0(x) \equiv 0$; nevertheless, the case $v_0(x) \leq 0$ is also relevant in the so called *suction problems*. To study problem (PS), several reformulations are proposed in the literature. The key point is to work with the *stream function* ψ given by

$$\begin{cases} u = \frac{\partial \psi}{\partial y} \\ v = -\frac{\partial \psi}{\partial x} + v_0, \ \psi(x,0) = 0 \end{cases}$$

Notice that the level lines of ψ coincide with the *current lines* of $\mathbf{v}$ = (u,v). The first mathematical treatment of (PS) is carried out by studying the third order ordinary differential equation satisfied by ψ (see Schlichting [27] for the case of Newtonian flows). The second possibility is to introduce the *von Mises transformation* [34]

$$\begin{array}{ll} \psi = \psi(x,y) & \psi \in (0,\infty), \\ w(x,\psi) \doteq u^2(x,y) & x \in (0,X). \end{array}$$

In this way, we arrive at the scalar problem

$$(P_w)\begin{cases} \frac{\partial w}{\partial x} - \nu\sqrt{w}\frac{\partial}{\partial \psi}\left(\left|\frac{\partial w}{\partial \psi}\right|^{p-2}\frac{\partial w}{\partial \psi}\right) + v_0\frac{\partial w}{\partial \psi} - 2UU_x = 0 & x \in (0,X), \psi \in (0,\infty), \\ w(0,\psi) = w_0(\psi) & \psi \in (0,\infty), \\ w(x,0) = 0 & x \in (0,X), \\ w(x,\psi) \to U^2(x) & as\ \psi \to \infty, \end{cases}$$

where $w_0(\psi)$ is defined through $u_0(y)$. The P.D.E. appearing in (P_w) is a nonlinear degenerate parabolic equation in which the x variable plays the role of *time* and ψ stands for the spatial variable. Some existence and uniqueness results for this problem are due to Oleinik [18], [19] (case of $p = 2$) and Samokhin [26] (case of $p > 2$). The assumptions of those papers are the following:

$$\begin{aligned} &U(x) > 0 && \textit{for } x \in (0, X), \\ &u_0(0) = 0 \textit{ and } u_0(y) > 0 && \textit{for } y > 0, \\ &u_0'(0) = 0\,,\ (u_0', u_0'') \in L^\infty(0,\infty)^2 \\ &U(0)U_x(0) - v_0(0)\tfrac{du_0}{dy} + \nu \left|\tfrac{du}{dy}\right|^{p-2} \tfrac{d^2u}{dy^2} = 0(y^2) && (\textit{consistency condition}). \end{aligned}$$

We also mention the results by Oleinik [19], Serrin [29] and Peletier [21] on the asymptotic behavior when X=+∞.

2 The results

The main goal of this work is to study the *coincidence set*

$$\{(x,\psi) : w(x,\psi) = U^2(x)\}$$

for the case of dilatant fluids. The boundary of this region could be called the *exact boundary layer*.

Remark 1 By the weak maximum principle, it is well known that necessarily $w(x,\psi) \leq U^2(x)$ in $(0,X) \times (0,\infty)$. In fact, if $p = 2$, it can be shown (see Oleinik [19]) that the strong maximum principle also holds and thus $w(x,\psi) < U^2(x)$ in $(0,X) \times (0,\infty)$, i. e. the coincidence set is empty. We recall that there are several attempts to make the boundary layer concept more precise. For instance, in Schlichting [27] it is defined as the zone where u=0. 99U. We must mention also the integral method introduced by von Karman [33] in order to estimate the boundary layer thickness δ.

Our main results are the following

Theorem 1 (Existence of the coincidence set).
Assume $p>2$, $v_0(x) \leq 0$, *and there exists* $\psi_0 \in (0,\infty)$ *such that* $w_0(\psi) = U^2(0)$ *for any* $\psi \geq \psi_0$; *Then there exists* $C > 0$ *and* $\mu \in (0,1)$ *such that*

$$w(x,\psi) = U^2(x) \textit{ for any } (x,\psi) \textit{ such that } \psi \geq \psi_0 + Cx^\mu.$$

Theorem 2 (Waiting distance along a streamline)
Assume $p > 2$, $v_0(x) \leq 0$ *and that there exists* $\psi_0 \geq 0$, $C > 0$ *and* $\sigma \in (0,1)$ *such that,*

$$\int_\psi^\infty (U^2(0) - w_0(\tau))^2 d\tau \leq C(\psi_0 - \psi)_+^{\sigma/(1-\sigma)}$$
$$\textit{for any } \psi \in (\psi_0 - \varepsilon, +\infty).$$

Then, there exists $x_0 \in (0,X)$ *such that*

$$w(x,\psi_0) = U^2(x) \textit{ for any } x \in [0,x_0].$$

Sketch of the proof of Theorem 1. It is based on a general *Energy Method* first introduced by one of the authors [1] and later improved and developed in [2], [3], [4], [5], [6], [7], [8], [9], [11], [12], [13], [14], [15], [16], [20] and [30] (see also [10], [23], [24], [31] and [32]).
First step. We introduce the homogenized unknown

$$z(x,\psi) \doteq U^2(x) - w(x,\psi).$$

We remark that by the comparison principle $z(x,\psi) \geq 0$ on $(0,X)\times(0,\infty)$. We also point out that arguing as in [18], or [26], it is possible to obtain the *a priori estimate*

$$0 < C_1 \leq U^2(x) - z(x,\psi) \leq C_2 \text{ for any } x \in (0,X), \psi \in (0,\infty),$$

for some constants $C_1 \leq C_2$. On the other hand, it is easy to see that z satisfies

$$(P_z)\begin{cases} \frac{\partial z}{\partial x} - \nu\sqrt{U^2 - z}\,\frac{\partial}{\partial\psi}\left(\left|\frac{\partial z}{\partial \psi}\right|^{p-2}\frac{\partial z}{\partial \psi}\right) + v_0\frac{\partial z}{\partial \psi} = 0 & x \in (0,X), \psi \in (0,\infty), \\ z(0,\psi) = U^2(0) - w_0(\psi) & \psi \in (0,\infty), \\ z(x,0) = U^2(x) & x \in (0,X), \\ z(x,\psi) \to 0 & as\ \psi \to \infty. \end{cases}$$

We remark that $z(x,\psi) = 0$ on the coincidence set.
Second step: *Integration by parts formula.* We introduce the one-parameter energy domain

$$Q_\rho = (0,x)\times(\rho,\infty)$$

where $\rho \geq \psi_0$ is arbitrary. Multiplying by z and by integrating (formally) by parts we obtain that

$$\begin{aligned} &\tfrac{1}{2}\int_\rho^\infty z^2(x,\psi)d\psi + \int_0^x\int_\rho^\infty \theta(s,\psi)\left|\tfrac{\partial z}{\partial \psi}(s,\psi)\right|^p dsd\psi + \\ &\tfrac{1}{2}\int_0^x(-v_0(s)z^2(s,\rho)ds = \tfrac{1}{2}\int_\rho^\infty z^2(0,\psi)d\psi - \int_0^x \sqrt{w}z\left|\tfrac{\partial z}{\partial \psi}\right|^{p-2}\tfrac{\partial z}{\partial \psi}\,|_{\psi=\rho}\, ds \end{aligned}$$

where

$$\theta(x,\psi) \doteq \frac{2U^2(x) - 3z(x,\psi)}{2\sqrt{U^2(x) - z(x,\psi)}}.$$

It is easy to see that

$$0 < C_3 \leq \theta(x,\psi) \leq C_4 \text{ for any } x \in (0,X), \psi \in (0,\infty).$$

Third step. We introduce the *energy functions*

$$b(x,\rho) \doteq ess \sup_{0\leq s\leq x} \frac{1}{2}\int_\rho^\infty z^2(x,\psi)d\psi$$

$$E(x,\rho) \doteq \int_0^x \int_\rho^\infty \theta(s,\psi) \left| \frac{\partial z}{\partial \psi}(s,\psi) \right|^p ds d\psi.$$

Applying the Holder inequality, we get that

$$\left| \int_0^x \sqrt{w} z \left| \frac{\partial z}{\partial \psi} \right|^{p-2} \frac{\partial z}{\partial \psi} |_{\psi=\rho} \, ds \right| \leq C_5 (-\frac{\partial E}{\partial \rho}(x,\rho))^{\frac{p-1}{p}} (\int_0^x |z(s,\psi)|^p \, ds)^{\frac{1}{p}}.$$

Now we need a technical result

Lemma 3 (Trace-interpolation inequality, [16]). *Let $\sigma \in (0,1)$ be given by $\sigma = (p+2)/3p$. Then*

$$(\int_0^x |z(s,\psi)|^p \, ds)^{\frac{1}{p}} \leq C_6 x^{(1-\sigma)/p} (E^{1/p} + x^{1/p} b^{1/2})^\sigma b^{(1-\sigma)/2}.$$

End of the proof of Theorem 1. By using the above inequalities we can find an exponent $\mu \in (0,1)$ and a positive constant C_7 such that

$$E^\mu \leq (E+b)^\mu \leq C_7 x^{(1-\mu)/(p-1)} (-\frac{\partial E}{\partial \rho}(x,\rho)).$$

This inequality implies the conclusion due to the following easy result

Lemma 4 ([1]) *Let $y \in C([0,t_1] \times [0,\rho_0+\delta]), y \geq 0$ such that*

$$\Phi(y(t,\rho)) \leq C t^\omega \frac{\partial y}{\partial \rho}(t,\rho)$$

for a. e. $\rho \in [0,\rho_0+\delta]$ and for any $t \in [0,t_1]$,where $\omega \geq 0$ and Φ is a continuous nondecreasing function such that $\Phi(0) = 0$ and

$$\int_{0+} \frac{ds}{\Phi(s)} < \infty.$$

Then there exists $t_0 \in (0,t_1]$ and a function $\rho(t)$ with $0 \leq \rho(t) \leq \rho_0 + \delta$ such that $y(t,\rho) = 0$ for any $t \in [0,t_0]$ and any $\rho \in [0,\rho(t)]$.

Remark 2 A different proof of Theorem 1, based upon the comparison principle, and under additional conditions, is due to [26].

Idea of the proof of Theorem 2. Using the same type of arguments and the assumption at $x = 0$ we obtain the differential inequality

$$E^\mu \leq C_7 x^{(1-\mu)/(p-1)} (-\frac{\partial E}{\partial \rho}(x,\rho)) + C_8 (\psi_0 - \psi)_+^{\mu/(1-\mu)}$$

for any $\psi \in (\psi_0 - \varepsilon, +\infty)$. The conclusion comes now from an analysis of this differential inequality

Lemma 5 ([2]) *Let* $y \in C([0, t_1] \times [0, \rho_0 + \delta]), y \geq 0$ *such that*

$$\Phi(y(t, s)) \leq Ct^{\omega}\frac{\partial y}{\partial \rho}(t, s) + G((\rho - \rho_0)_+)$$

for a. e. $\rho \in [0, \rho_0 + \delta]$ *and for any* $t \in [0, t_1]$, *where* $\omega \geq 0$ *and* Φ *is as in Lemma 4 and*

$$\exists \mu > 0 \text{ and } \varepsilon > 0 \text{ such that } G(s) \leq \varepsilon\Phi(\eta_\mu(s)), \text{ a.e. } s \in (0, \rho)$$

with

$$\eta_\mu(s) = \Theta_\mu^{-1}(s), \qquad \Theta_\mu(\tau) = \int_{0+}^{\tau} \frac{ds}{\mu\Phi(s)}.$$

Then there exists $t^* \in (0, t_1]$ *such that* $y(t, \rho) = 0$ *for any* $t \in [0, t^*]$ *and any* $\rho \in [0, \rho_0]$.

Remark 3 In the case of pseudo-plastic fluids ($1 < p < 2$), it is possible to apply another kind of energy method (now using a suitable global energy) which leads to a different estimate on the location of the exact boundary layer: if X is large enough and U(x)$\equiv$ 0 $\forall x \geq x_U$ for some $x_U > 0$, then there exists $x_0 \geq x_U$ such that $w(x, \psi) = U^2(x) = 0$, $\forall x \geq x_0$, $\forall \psi \in (0, \infty)$ (see also [25]).

Acknowledgments. *The research of the authors is partially sponsored by the project INTAS-94-2187. The second author is also supported by the DGICYT (Spain) project PB93-0443.*

References

1. Antontsev, S.N.: On the localization of solutions of nonlinear degenerate elliptic and parabolic equations. *Dokl. Akad. Nauk. SSSR*, 1981, 260, No. 6, pp. 1289-1293. (in Russian; English translation in *Soviet Math. Dokl.* , 1981, 24, No. 2, pp. 420-424).

2. Antontsev, S.N. and Diaz, J.I.: Application of an Energy Method in the Localization of the Solutions of the Equations of Continuous medium mechanics. (Russian). *Dokl. Akad. Nauk. SSSR.* 1988. Vol 303. No 2. pp. 320 - 325. (English transl. in *Soviet Phys. Dokl.* 1988. Vol 33. No 11. pp. 813 - 816.).

3. Antontsev, S.N. and Diaz, J.I.: New results on localization of solutions of nonlinear elliptic and parabolic equations obtained by the energy method. *Dokl. Akad. Nauk. SSSR.* 1988, Vol. 303, No. 3, pp. 320-325. (English translation in *Soviet Math. Dokl.* 1988, Vol. 33, No. 3,pp. 535-539).

4. Antontsev, S.N. and Diaz, J.I.: The Energy Method and the Localization of Solutions of Equations in Continuum Mechanics. (Russian). *Zh. Prikl. Mekh. i. Tekhn Fiz.* 1989. No 2. pp. 18-25. (English transl. in *J. Appl. Mech. Tech. Phys.* 1989. Vol 30. No 2. pp. 182-189).

5. Antontsev, S.N. and Diaz, J.I.: On Space or Time Localization of Solutions of Nonlinear Elliptic or Parabolic Equations Via Energy Methods. (English). In *Recent Advances In Nonlinear Elliptic and Parabolic Problems*, Ph. Benilan, L.C. Evans and M. Pierre eds. Pitman Research Notes In Mathematics. Series 208, 1989, pp. 3-14.

6. Antontsev, S.N. and Diaz, J.I.: Space and Time Localization in the Flow of Two Immiscible Fluids Through a Porous medium: Energy Method Applied to Systems. *Nonlinear Analysis, Theory, Methods and Applications*, 1991, Vol 16. No 4. pp. 299-313.

7. Antontsev, S.N., Diaz, J.I. and Shmarev, S.I.: New applications of energy methods to parabolic free boundary Problems. *Uspekhi Mat. Nauk.*, 1991, 46, No. 6 (282), pp. 181-182. (in Russian; English translation: in *Russian Mathematical Surveys*, 1991, 46, No. 6, pp. 193-194).

8. Antontsev, S.N., Diaz, J.I. and Shmarev, S.I.: The support shrinking properties for solutions of quasilinear parabolic equations with strong absorption terms. *Annales de la Fac. des Sciences de Toulouse,* 1995, 4, No. 1, pp. 5-30.

9. Antontsev, S.N., Shmarev, S.I.: Local energy method and vanishing of weak solutions of nonlinear parabolic equations. *Dokl. Akad. Nauk. SSSR*, 1991, 318, No. 4. pp. 777-781. (in Russian; English translation in *Soviet Math. Doklady*, 1991, No. 43).

10. Benilan, Ph. and Crandall, M.G.: The continuous dependence on φ of the solutions of $u_t - \Delta\varphi(u) = 0$, *Indiana Univ. Math. J.*, No. 30, 1981, pp. 161-177.

11. Bernis, F.: Compactness of the support for some nonlinear elliptic problems of arbitrary order in dimension n, *Comm. Partial Differential Equations* 1984, 9, pp. 271-312.

12. Bernis, F.: Finite speed of propagation and asymptotic rates for some nonlinear higher order parabolic equations with absorption, *Proc. Royal Soc. Edinburgh 104A*, 1986, pp. 1-19.

13. Bernis, F.: Qualitative properties for some nonlinear higher order degenerate parabolic equations, *Houston J. Math.*, 14, 1988, pp. 319-352.

14. Diaz, J.I.: *Nonlinear partial differential equations and free boundaries.* Research Notes in Math. No. 106. Pitman, 1985.

15. Diaz, J.I. and Liñan, A.: Movimiento de descarga de gases en conductos largos: Modelizacion y estudio de una ecuacion parabolica doblemente no lineal. *Actas de la Reunion Matematica en Honor de A. Dou*, J.I. Diaz and J.M. Vegas eds., Editorial Universidad Complutense, Madrid, 1989, pp. 95-119.

16. Diaz, J.I. and Veron, L.: Local vanishing properties of solutions of elliptic and parabolic quasilinear equations. *Trans. Amer. Math. Soc.*, 1985, 290, No. 2, pp. 787-814.

17. Narasimhan, M.N.L.: *Principles of Continuum Mechanics*, John Wiley and Sons, Inc. New York, 1993.

18. Oleinik, O. A.: On a system of equations in boundary layer theory, *Zhurn. Vyvhslit. Mat. Mat. Fiz.* 1963, 3, pp. 489-507 [Russian]. Engl. transl. USSR *Comput. Math. Math. Phys.* 1963, 3, pp. 650-703.

19. Oleinik O. A.: *Mathematical problems of boundary layer theory*. Minneapolis. Univ. of Minnesota. 1969.

20. Palymsky I. B.: Some qualitative properties of the solutions of equations of nonlinear heat conductivity with absorption. *Chislennye metody mekhaniki sploshnoi sredy*, 1985, Vol 16. No. 1, pp. 136-145.

21. Peletier, L.A.: On the asymptotic behavior of velocity profiles in laminar boundary layers, *Arch. Rat. Mech. Anal.* 45, 1972, pp. 110-119.

22. Prandtl, L.: Über Flussigkeits-be wegungen bei sehr kleiner Reibung. In *Verh. II. Int. Math-Kongr. Heidelberg*, 1904, pp. 484-494. Leipzig: Teubner.

23. Rykov, Yu.G.: On the behavior of generalized solutions of degenerating parabolic equations with anisotropic non-power nonlinearities with the growth of time. *Differential Equations and Their Applications*. Moscow. Gos. Univ. Moscow. 1984, pp. 160-164.

24. Rykov, Yu.G.: On the finite velocity of propagation of perturbations and inertia for generalized solutions of degenerating quasilinear equations. *Differential Equations and Their Applications*. Moscow. Gos. Univ. Moscow. 1984, pp. 160-164.

25. Samokhin, V.N.: Generalized solutions and extension of the boundary layers of a pseudo-plastic liquid. *Trudy. Sem. im. I. G. Petrovskogo*, No. 3, 1978, pp. 161-175.

26. Samokhin, V.N.: A system of steady–state boundary layer for a dilatantous medium. *Trudy Seminara imeni I. G. Petrovskogo*. No. 14, 1989, pp. 89-109. (English transl. 1990, pp. 2358-2373).

27. Schlichting, H.: *Boundary-Layer Theory*, McGraw-Hill Company, New York, 7th edition, 1979

28. Schowalter, R.: *Mechanics of non-Newtonian Fluids*, Pergamon Press, Oxford, 1978.

29. Serrin, J.: On the mathematical basis for Prandtl's boundary layer theory: An example. *Arch. Rat. Mech. Anal.* 28, 1967, pp. 217-225.

30. Shmarev, S.I.: Local properties of solutions of nonlinear equation of heat conduction with absoption. *Dinamika Sploshnoi Sredy*, Novosibirsk, 1990, 95, pp. 28-42. (in Russian).

31. Straughan, B.: *Instability, nonexistence and weighted energy methods in fluid dynamics and related theories*. Pitman Advanced Publishing Program. Research Notes in Math., 74, 1982.

32. Straughan, B.: *The energy method, stability and nonlinear convection*. Applied Mathematical. Sciences. 1991. Springer Verlag.

33. von Karman, Th.: Über laminare und turbulente Reibung, *Z. Angew. Math. Mech.* 1 1927, pp. 233-252.

34. von Mises, R.: Bemerkungen zur Hydrodynamic, *Z. Angew. Math. Mech.* 7 1927, pp. 425-431.

MINIMAL ENERGY FOR A FREE BALL ON AN ELASTIC MEMBRANE

J. BEMELMANS
Institut für Mathematic, RWTH Aachen, Templergraben 55
D-52062 Aachen (Germany),

M. CHIPOT
Centre d'Analyse Non Linéaire, Université de Metz, URA-CNRS 399, Ile de Sauley, 57045 Metz-codex 01 (France)

1. Introduction

Let us denote by Ω a bounded domain of $\mathbf{R}^2$ with smooth boundary Γ. Ω represents an elastic membrane spanned over a frame Γ.

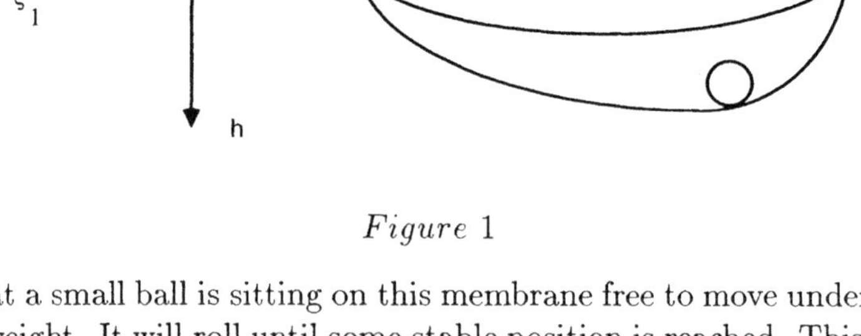

Figure 1

Assume that a small ball is sitting on this membrane free to move under the action of its own weight. It will roll until some stable position is reached. This is what we would like to analyse here i.e. we would like to determine what conditions insure

S. N. Antontsev et al. (eds.), Energy Methods in Continuum Mechanics, 22–30.

the existence of such an equilibrium position for the heavy ball. The problem has been addressed before from a different point of view in [EF], see also [G], [BF].

Let us denote by G the weight of the ball and by r its radius. Its center will be labelled by $(\zeta_1, \zeta_2, h) = (\zeta, h)$ -note on *Figure* 1 the orientation chosen for the h-axis. If u denotes the vertical displacement of the membrane when the ball has its center located at (ζ, h) the energy of this configuration is given by

$$\mathcal{E}(u, \zeta, h) = \frac{1}{2}\int_\Omega |\nabla u(x)|^2\, dx - Gh. \tag{1.1}$$

Moreover, the ball remaining on one side of the membrane, u has to satisfy the constraints

$$u(x) = 0 \quad \text{on} \quad \Gamma \quad , \quad u(x) \geq h + \sqrt{r^2 - |x-\zeta|^2} \quad \text{on} \quad B_r(\zeta) \tag{1.2}$$

where $B_r(\zeta)$ denotes the ball of center ζ and radius r. We will set in the sequel

$$\Psi(x) = \Psi_{\zeta,h}(x) = h + \sqrt{r^2 - |x-\zeta|^2} \quad \forall\ x \in \ B_r(\zeta). \tag{1.3}$$

Now, we will consider only positions of the center for which the ball is not touching the rim of the membrane i.e. if $dist(.,\Gamma)$ denotes the distance from a point to Γ we will assume that ζ satisfies

$$dist(\zeta, \Gamma) > r. \tag{1.4}$$

Then, setting

$$\Omega_r = \{\zeta \in \Omega \mid dist(\zeta,\Gamma) > r\} \tag{1.5}$$

we would like to find out if there exists (u_0, ζ_0, h_0) such that u_0 is admissible i.e.

$$\zeta_0 \in \Omega_r \quad , \quad u_0(x) = 0 \ \text{ on } \ \Gamma \quad , \quad u_0(x) \geq \Psi_{\zeta_0,h_0}(x) \ \text{ on } B_r(\zeta_0) \tag{1.6}$$

and

$$\mathcal{E}(u_0, \zeta_0, h_0) = Inf\{\mathcal{E}(u,\zeta,h) \mid \zeta \in \Omega_r,\ u = 0 \ \text{ on } \ \Gamma,\ u \geq \Psi_{\zeta,h} \ \text{ on } B_r(\zeta)\}. \tag{1.7}$$

In the framework of the Sobolev spaces let us introduce

$$K_{\zeta,h} = \{v \in H_0^1(\Omega) \mid v(x) \geq \Psi_{\zeta,h}(x) \text{ a.e. } x \in \ B_r(\zeta)\} \tag{1.8}$$

and denote by $u = u_{\zeta,h}$ the solution to the obstacle problem

$$\min_{v \in K_{\zeta,h}} \frac{1}{2}\int_\Omega |\nabla v(x)|^2\, dx - Gh \tag{1.9}$$

i.e. u satisfies the Variational Inequality (see [KS])

$$\begin{cases} u \in K_{\zeta,h}, \\ \int_\Omega \nabla u(x)\nabla(v-u)(x)\,dx \geq 0 \quad \forall\, v \in K_{\zeta,h}. \end{cases} \tag{1.10}$$

Then, it is clear that the problem can be reformulated as the search of $(\zeta_0, h_0) \in \Omega_r \times \mathbf{R}$ such that $u_0 = u_{\zeta_0,h_0}$ satisfies

$$\frac{1}{2}\int_\Omega |\nabla u_{\zeta_0,h_0}(x)|^2\,dx - Gh_0 =$$

$$min\{\frac{1}{2}\int_\Omega |\nabla u_{\zeta,h}(x)|^2\,dx - Gh \mid (\zeta,h) \in \Omega_r \times \mathbf{R}\} \tag{1.11}$$

i.e. the problem reduces to find (ζ_0, h_0) realizing the infimum of the function

$$F(\zeta,h) = \frac{1}{2}\int_\Omega |\nabla u_{\zeta,h}(x)|^2\,dx - Gh \tag{1.12}$$

over the set of $(\zeta,h) \in \Omega_r \times \mathbf{R}$. If it is relatively easy to show that F is a continuous function of its arguments, one should note that $\Omega_r \times \mathbf{R}$ is neither closed nor compact.

Remark 1.1 : One could also consider the problem of a small ball on a soap film spanned over Γ. In this case the situation is slightly different and one has to replace the Dirichlet integral by

$$\int_\Omega \sqrt{1 + |\nabla u(x)|^2}\,dx. \tag{1.13}$$

Of course, due to the fact that the soap film cannot support too heavy loads, in this case some limitation on the range of h has to be introduced. We refer the interested reader to [BC] for details.

2. Existence of a solution

In order to recast the problem on a compact set let us first prove :

Lemma 2.1 : One has

$$\lim_{h \to \pm\infty} F(\zeta,h) = +\infty \tag{2.1}$$

uniformly in $\zeta \in \Omega_r$.

proof : When $h \to -\infty$ this is due to the fact that

$$F(\zeta, h) = \frac{1}{2} \int_\Omega |\nabla u_{\zeta,h}(x)|^2 \, dx - Gh \geq -Gh. \tag{2.2}$$

When now $h \to +\infty$, remark that by the Poincaré Inequality one has for some constant C

$$\begin{aligned} F(\zeta, h) &\geq C \int_\Omega |u_{\zeta,h}(x)|^2 \, dx - Gh \\ &\geq C \int_{B_r(\zeta)} |\Psi_{\zeta,h}(x)|^2 \, dx - Gh \\ &\geq C|B_r(\zeta)|h^2 - Gh \end{aligned} \tag{2.3}$$

where $| \ |$ denotes the usual Lebesgue measure. This clearly completes the proof of the Lemma.

Next we can show :

Theorem 2.1 : Assume h fixed. Then the function

$$\zeta \to D(\zeta) = \frac{1}{2} \int_\Omega |\nabla u_{\zeta,h}(x)|^2 \, dx \tag{2.4}$$

is differentiable in Ω_r and for any unit vector ν one has

$$\frac{\partial D}{\partial \nu}(\zeta) = - \int_\Lambda \frac{\partial \Psi}{\partial \nu}(x) \Delta(x) \, dx \tag{2.5}$$

where Λ denotes the coincidence set of u i.e.

$$\Lambda = \{x \in B_r(\zeta) \mid u(x) = \Psi(x)\}. \tag{2.6}$$

Proof : Let us give here a formal proof. The reader is refered to [BC] for a rigorous derivation of the arguments below. First remark that by differentiation under the integral sign one has

$$\begin{aligned} \frac{\partial}{\partial \nu}\{\frac{1}{2} \int_\Omega |\nabla u(x)|^2 \, dx\} &= \int_\Omega \nabla u \cdot \nabla(\frac{\partial u}{\partial \nu}) \, dx \\ &= \int_\Omega \nabla \cdot (\nabla u \frac{\partial u}{\partial \nu}) \, dx - \int_\Omega \Delta u \frac{\partial u}{\partial \nu} \, dx \\ &= \int_\Gamma \frac{\partial u}{\partial n} \frac{\partial u}{\partial \nu} \, d\sigma(x) - \int_\Omega \Delta u \frac{\partial u}{\partial \nu} \, dx \end{aligned} \tag{2.7}$$

where $d\sigma(x)$ denotes the superficial measure on Γ, n the outward unit normal to Γ. Now, on Γ one has for every ζ

$$u = u_{\zeta,h} \equiv 0 \tag{2.8}$$

so that

$$\frac{\partial u}{\partial \nu} = 0. \tag{2.9}$$

It follows then from (2.7) that

$$\begin{aligned} \frac{\partial}{\partial \nu}\{\frac{1}{2}\int_\Omega |\nabla u(x)|^2\, dx\} &= -\int_\Omega \Delta u \frac{\partial u}{\partial \nu}\, dx \\ &= -\int_\Lambda \Delta\Psi \frac{\partial \Psi}{\partial \nu}\, dx \end{aligned} \tag{2.10}$$

since $\Delta u = 0$ outside of Λ (see [KS]).

Having proved this formula we would like to show how it can be used to establish that the free energy of he ball decreases when one gets away from the boundary Γ. The situation is described in the figure below.

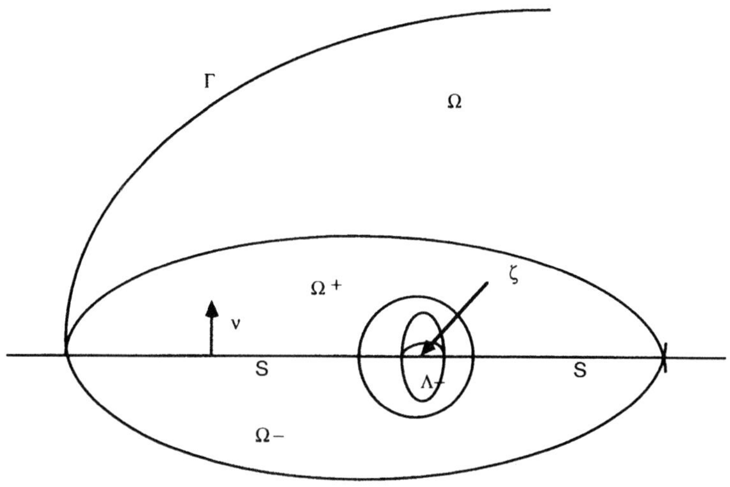

Figure 2

Theorem 2.2 : Let $\zeta \in \Omega_r$ and ν a unit vector. Let R denote the reflection in the axis S (see *Figure* 2) going through ζ and orthogonal to ν. Assume that $h > -r$ and set

$$\Omega^- = \{x \in \Omega \mid (x - \zeta \cdot \nu) < 0\} \tag{2.11}$$

$$\Omega^+ = \{x \in \Omega \mid (x - \zeta \cdot \nu) > 0\} \tag{2.12}$$

$$\Lambda^- = \Lambda \cap \Omega^- \quad , \quad \Lambda^+ = \Lambda \cap \Omega^+. \tag{2.13}$$

If

$$R(\Omega^-) \subset \Omega^+ \tag{2.14}$$

the inclusion being strict, then

$$\Lambda^+ \subset R(\Lambda^-) \tag{2.15}$$

and the inclusion is strict. Moreover,

$$\frac{\partial D}{\partial \nu}(\zeta) < 0. \tag{2.16}$$

Proof : The inclusion (2.15) relies on a careful application of the maximum principle. Now, (2.16) is based on the fact that the coincidence set is unbalanced and (2.5). Indeed, by (2.5) one has

$$\frac{\partial D}{\partial \nu}(\zeta) = -\int_\Lambda \frac{\partial \Psi}{\partial \nu} \Delta\Psi \, dx =$$

$$-\int_{\Lambda+} \frac{\partial \Psi}{\partial \nu} \Delta\Psi \, dx - \int_{R(\Lambda+)} \frac{\partial \Psi}{\partial \nu} \Delta\Psi \, dx - \int_{\Lambda^- \setminus R(\Lambda+)} \frac{\partial \Psi}{\partial \nu} \Delta\Psi \, dx. \tag{2.17}$$

Since Ψ is radially symmetric a simple computation shows that

$$\Delta\Psi < 0$$

on $B_r(\zeta)$ and of course $\Delta\Psi$ is radially symmetric. Moreover, one has clearly

$$\frac{\partial \Psi}{\partial \nu}(x) = \frac{1}{\sqrt{r^2 - |x - \zeta|^2}} \, (x - \zeta \cdot \nu) \tag{2.18}$$

so that

$$\frac{\partial \Psi}{\partial \nu}(Rx) = -\frac{\partial \Psi}{\partial \nu}(x).$$

Going back to (2.17) we obtain

$$\int_{\Lambda+} \frac{\partial \Psi}{\partial \nu} \Delta\Psi \, dx = -\int_{R(\Lambda+)} \frac{\partial \Psi}{\partial \nu} \Delta\Psi \, dx$$

and thus

$$\frac{\partial D}{\partial \nu}(\zeta) = -\int_{\Lambda^- \setminus R(\Lambda+)} \frac{\partial \Psi}{\partial \nu} \Delta\Psi \, dx < 0 \tag{2.19}$$

which gives (2.16). It is clear that the above theorem allows us to derive an existence result for the problem (1.7), (1.11). Indeed one has :

Theorem 2.3 : Assume that for any r'

$$r < r' < r + \epsilon$$

and any $\zeta \in \Omega$ such that

$$dist(\zeta, \Gamma) = r' \tag{2.20}$$

there exists a vector ν such (2.14) holds, then the problem (1.11) admits a minimizer.

Proof : Applying Theorem 2.2 we see that the infimum of F cannot be achieved at a point where (2.20) holds. So, combining this remark with Lemma 2.1 we see that F attains its minimum on the compact set

$$\{\zeta \in \Omega \mid dist(\zeta, \Gamma) \geq r + \epsilon\} \times [-h^*, h^*] \tag{2.21}$$

where h^* is some positive number. This completes the proof of the theorem provided we have shown that F is continuous (see [BC]).

Remark 2.1 : It is clear, see Theorem 2.2, that when we are under the assumptions of Theorem 2.3 and when Ω admits an axis of symmetry, then the infimum of F has to be achieved at a point ζ of this axis. When Ω admits two axis of symmetry the infimum has to be achieved at their intersection.

In the case where Ω is a disk one can compute exactly the solution to (1.7) or (1.11). More precisely if

$$\Omega = B_R(0)$$

the ball of center 0 and radius R then the coincidence set Λ of the solution u_0 is given by

$$\Lambda = B_{r_0}(0) \tag{2.22}$$

with r_0 solution to

$$\frac{r_0^2}{\sqrt{r^2 - r_0^2}} = \frac{G}{2\pi}. \tag{2.23}$$

Moreover,

$$u_0(x) = \begin{cases} \Psi = h_0 + \sqrt{r^2 - |x|^2} & \text{for } |x| \leq r_0, \\ -\frac{G}{2\pi} ln\frac{|x|}{R} & \text{for } |x| \geq r_0, \end{cases} \tag{2.24}$$

with h_0 given by

$$h_0 = -\frac{G}{2\pi} ln\frac{r_0}{R} - \sqrt{r^2 - r_0^2}. \tag{2.25}$$

We refer the reader to [BC] for details. Remark that

$$h_0 = h_0(R) \to +\infty \tag{2.26}$$

when $R \to +\infty$.

3. Uniqueness

In the case where Ω is convex and admits a single axis of symmetry, even if uniqueness is likely we do not know how to prove it in general (see [EF], [G] for the limit case $r \to 0$). In the case where Ω fails to be convex we can show however that uniqueness can fail as well. More precisely consider the case of *Figure* 3 where

$$\Omega = \Omega_R = (-S, S) \times (-1, 1) \cup B_R(-S) \cup B_R(S) \quad , \quad S = R + 1. \tag{3.1}$$

$B_R(-S)$, $B_R(S)$ denote respectively the balls of center $(-S, 0)$, $(S, 0)$ and radius R (see below).

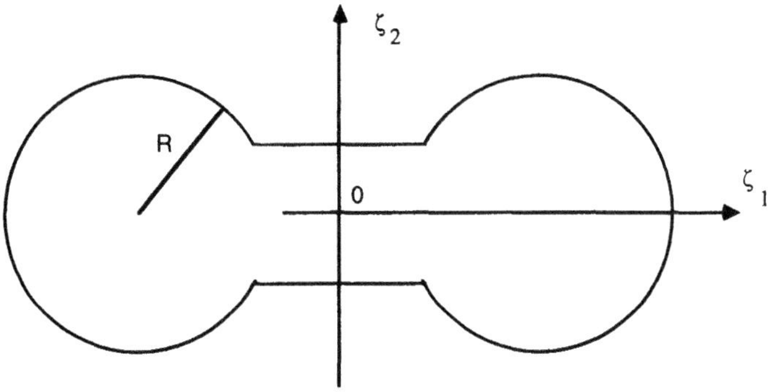

Figure 3

Let us assume that $r < 1$. Moreover, let us denote by E the infimum of the energy $\mathcal{E}$, (see 1.1), when $\Omega = Q = (-1, 1)^2$ and suppose that the problem (1.7), (1.11) admits a unique minimizer u_R when $\Omega = \Omega_R$. For symmetry reasons the corresponding ζ_0 must be equal to 0. Let us denote by $h_0 = h_0(R)$ the corresponding height of the ball. Clearly, since $Q \subset \Omega_R$

$$E \geq \mathcal{E}(u_R, 0, h_0)$$

$$= \frac{1}{2}\int_{\Omega_R} |\nabla u_R|^2\, dx - Gh_0$$

$$\geq \frac{1}{2}\int_{Q} |\nabla u_R|^2\, dx - Gh_0(R)$$

$$\geq -Gh_0(R) \tag{3.2}$$

when $h_0 \leq 0$. When $h_0 > 0$, using the Poincaré Inequality on Q (note that $u_R = 0$ on some part of its boundary) we obtain

$$E \geq \frac{1}{2}\int_{Q} |u_R|^2\, dx - Gh_0(R)$$

$$\geq \frac{1}{2}|B_r(0)|h_0^2(R) - Gh_0(R). \tag{3.3}$$

Since E is clearly bounded independently of R, so does $h_0(R)$ by (3.2) and (3.3) and for some constant M one has

$$\mathcal{E}(u_R, 0, h_0) \geq M. \tag{3.4}$$

But going back to (2.22)-(2.25), it is easy to see that the energy for a minimizer of $\mathcal{E}$ on $\Omega = B_R(0)$ tends to $-\infty$ when $R \to +\infty$. So, considering such a minimizer on $B_R(-S)$ or $B_R(S)$ and noting that this is an admissible function for the infimum (1.7) would lead to a contradiction to (3.4). So, for R large enough, the problem (1.7) corresponding to $\Omega = \Omega_R$ cannot have a unique minimizer.

References

[BF] Bandle, C. and Flucher, M. (1994) Harmonic radius and concentration of energy, hyperbolic radius and Liouville's equation $\Delta u = e^u$ and $\Delta u = u^{\frac{n+2}{n-2}}$. Preprint.

[BC] Bemelmans, J. and Chipot, M. (1995) On a variational problem for an elastic membrane supporting a heavy ball. Calculus of Variations and PDE, (to appear).

[EF] Elliot, C. M. and Friedman, A. (1986) The contact set of a rigid body partially supported by a membrane. Nonlinear Analysis, 10, 251–276.

[G] Gustaffson, B. (1990) On the convexity of a solution of Liouville's equation. Duke Math. J. 60, 303-311

[KS] Kinderlehrer, D. and Stampacchia, G. (1980) *An Introduction to Variational Inequalities and their Applications.* Academic Press, New York.

ENERGY METHODS FOR HIGHER ORDER ELLIPTIC AND PARABOLIC PROBLEMS

Francisco BERNIS
Departamento de Matemáticas, Universidad Autónoma de Madrid
28049 Madrid, Spain

1. Introduction

In these notes we take as a model from the expository point of view the fourth order elliptic equation

$$\Delta^2 u + |u|^a \operatorname{sgn} u = f(x) \quad \text{in} \quad \Omega \subset \mathbf{R}^n \tag{1}$$

and the parabolic equation

$$\frac{\partial}{\partial t}(|u|^a \operatorname{sgn} u) + \Delta^2 u = 0 \quad \text{in} \quad \Omega \times (0, \infty) \tag{2}$$

The equation (1) is related to elastic plate models, while the equation (2) is a sort of "fourth order porous medium equation".

The results and methods of these notes apply to a broad class of equations of higher order. For example, Δ^2 can be replaced by $(-\Delta)^m$ and even by A_p, a nonlinear p-Laplacian of order $2m$ such us

$$A_p u = (-1)^m \sum_{|\alpha|=m} D^\alpha(|D^\alpha u|^{p-1} \operatorname{sgn} D^\alpha u)$$

If $0 < a < 1$ the equation (1) has the *compact support property,* which means that the solutions of (1) have compact support if the data have compact support. Even when the domain is bounded and the boundary conditions are nonhomogeneous the solution may be zero in a region with nonempty interior and then it is said that there exists a *dead core.*

In the case of parabolic equations the compact support property is also known as *finite speed of propagation.* The equation (2) has finite speed of propagation if $0 < a < 1$: see [5].

These notes are mainly devoted to outline a proof of the compact support and dead core properties for Equation (1) by means of an energy method. Background references are [2, 3, 9].

S. N. Antontsev et al. (eds.), Energy Methods in Continuum Mechanics, 31–37.

2. The energy method

This is a broad concept. We just try to describe some aspects of our work [2-9] on energy methods. We quote [1, 10, 11] for similar methods for second order problems. The method has 3 ingredients.

1) The first ingredient is to multiply the equation by u and integrate by parts in some *variable* or moving subdomain, e.g. a variable half-space or a variable ball. This leads to "energy integrals" such as

$$\int |D^2u|^2\,dx \quad \text{or} \quad \int |D^2u|^2\,dx + \int |u|^{a+1}\,dx$$

plus other terms over the boundary of the subdomain. It may be better to multiply by ρu, where ρ is a *weight function*. Then one obtains the weight in the energy integrals. The weight is used to cancel the boundary terms, but it generates other terms (from derivatives of products). In second order problems the boundary terms can be handled directly and weights are not necessary for typical examples.

2) The second ingredient is the use of interpolation-imbedding inequalities. These inequalities are related to Sobolev imbedding inequalities, but with some complications and refinements. They need to incorporate the weights introduced before. In these notes we shall use a Hardy inequality, some weighted Gagliardo-Nirenberg inequalities and a Caccioppoli type inequality.

3) The third ingredient is an ordinary differential inequality involving energy integrals. The independent variable of this differential inequality is just the variable which labels the moving subdomain.

In the sequel we shall apply the energy method first using half-spaces and then using balls. The use of half-spaces is more transparent, but requires zero Dirichlet boundary conditions. When using balls the boundary conditions may not be zero.

As notational convention, C will stand for a positive constant which may be different in different occurrences.

3. The inequalities for the energy method in half-spaces

Notation: $x = (x_1, \cdots, x_n)$ $\quad dx = dx_1 \cdots dx_n$

Domain of integration: The half-space $\{x_n > 0\}$

Hardy inequality: If $\lambda > -1$

$$\int x_n^\lambda u^2\,dx \le \left(\frac{2}{\lambda+1}\right)^2 \int x_n^{\lambda+2}\left(\frac{\partial u}{\partial x_n}\right)^2 dx \tag{3}$$

This follows from the one-dimensional inequality [13, Theorem 330] plus Fubini's theorem: see e.g. [5, Appendix I].

A weighted Gagliardo-Nirenberg inequality

$$\left(\int x_n^\lambda |D^j u|^2\,dx\right)^{1/2} \le C\left(\int x_n^\lambda |D^2u|^2\,dx\right)^{\theta/2}\left(\int x_n^\lambda |u|^q\,dx\right)^{(1-\theta)/q} \tag{4}$$

if $j = 0$ or $j = 1$, $\lambda \geq 0$ and $0 < q \leq 2$
where θ satisfies $0 \leq \theta \leq 1$ and is given by

$$\frac{\lambda + n}{2} - j = \theta(\frac{\lambda + n}{2} - 2) + (1 - \theta)\frac{\lambda + n}{q}$$

The inequality (4) is taken from [2, Appendix I].

Remark 1. For $\lambda = 0$ the relation (4) is a Gagliardo-Nirenberg inequality [12] and [14]. Here it is presented as in [14].

Remark 2. If the domain is the half-space $\{x_n > z\}$ replace x_n^λ by $(x_n - z)^\lambda$. Of course the constant C does *not* change.

Remark 3. The inequalities (4) are related to imbedding inequalities for weighted Sobolev spaces, but they have the following special features: 1) They involve different powers, since q may be different from 2. 2) There are no first order derivatives in the right-hand side. 3) The product form of the right-hand side.

Remark 4. For bounded domains the "pure" product form of (4) is not true: an additional term is needed unless one assumes homogeneous boundary conditions. See for example the inequality (8) below.

4. Compact support property using half-spaces

We proceed to sketch the proof of the compact support property for Equation (1) using the energy method in half-spaces. We make formal computations whose justification can be found in [2] and [3]. We assume zero Dirichlet boundary conditions, i.e.

$$u = D^1 u = 0 \quad \text{on} \quad \partial\Omega$$

In this section the domain of integration is the intersection of Ω and the half-space $\{x_n > z\}$. We shall take $z \geq 0$.

Let $q = a + 1$, assume that support f and the half-space $\{x_n > 0\}$ do not intersect and set the notation

$$I_s(z) \equiv \int (x_n - z)^s |D^2 u|^2 \, dx + \int (x_n - z)^s |u|^q \, dx$$

We multiply Equation (1) by $(x_n - z)^4 u(x)$ and integrate on $\{x_n > z\}$. For the integrations by parts we use the identity

$$\Delta^2 = \sum_{|\alpha|=2} D^\alpha D^\alpha \quad \text{rather than} \quad \Delta^2 = \Delta\Delta$$

After some computations with the products

$$\sum_{|\alpha|=2} (D^\alpha u) \, D^\alpha((x_n - z)^4 u)$$

we obtain

$$I_4(z) \le C \int (x_n - z)^3 |D^2 u| \cdot |D^1 u| \, dx + C \int (x_n - z)^2 |D^2 u| \cdot |u| \, dx$$

which in turn is bounded by

$$C \left(\int (x_n - z)^4 |D^2 u|^2 \right)^{1/2} \left(\int (x_n - z)^2 |D^1 u|^2 \right)^{1/2}$$
$$+C \left(\int (x_n - z)^4 |D^2 u|^2 \right)^{1/2} \left(\int |u|^2 \right)^{1/2}$$

Absorbing in the left-hand side the derivatives of second order and using Hardy inequality (3) for the integral of u^2 it follows that

$$I_4(z) \le C \int (x_n - z)^2 |D^1 u|^2 \, dx$$

and then from (4) and the numerical inequality $A^a B^b \le C(A+B)^{a+b}$ we finally obtain

$$I_4(z) \le C I_2(z)^\mu \quad , \quad \mu > 1 \tag{5}$$

where the exponent μ can be explicitly computed from (4).

Notice that $I_4'(z) = -4I_3(z)$ and $I_4''(z) = 12I_2(z)$. From the *differential inequality* (5) it can be derived that $u = 0$ for x_n large enough.

Remarks on the method

1. The method applies to weak solutions: it amounts to the use of an admissible test function: see [3] for elliptic equations and [5] and [6] for degenerate parabolic equations.

2. The method applies to quasilinear equations involving some types of p-Laplacian operators of order $2m$.

3. The method gives estimates of the support (or of the free boundary) in terms of the energy (without weights) and, hence, in terms of the data.

5. Dead core property using balls

We proceed to sketch the proof of the dead core property for Equation (1) using the energy method in balls. Now the solution u may have nonhomogeneous boundary values.

Let B_R be a ball of radius R and a fixed center, $0 < R < R_0$. We assume that

$$B_{R_0} \subset \Omega \setminus \text{ support } f$$

In this section the domain of integration is always the ball B_R.

We introduce the notation (recall that $q = a + 1$):

$$J_s(R) \equiv \int (R - |x|)^s |D^2 u|^2 \, dx + \int (R - |x|)^s |u|^q \, dx$$

We need the following *Caccioppoli type inequality*, whose proof can be found in [8]. If $s \geq 2m$ then

$$\int (R-|x|)^s u\,(-\Delta)^m u\,dx \geq (1-\varepsilon)\int (R-|x|)^s |D^m u|^2\,dx - C\int (R-|x|)^{s-2m} u^2\,dx \tag{6}$$

where the constant C is independent of R.

Applying (6) to Equation (1) with $m=2$ and $s=4$ we obtain

$$J_4(R) \leq C\int u^2\,dx \tag{7}$$

Notice that $J_4'(R) = 4J_3(R)$.

Next we need the following Gagliardo-Nirenberg inequality (see [15]) for the ball B_R.

$$C\left(\int u^2\right)^{1/2} \leq \left(\int |D^2 u|^2\right)^{\theta/2}\left(\int |u|^q\right)^{(1-\theta)/q} + R^{-\beta}\left(\int |u|^q\right)^{1/q} \tag{8}$$

where C is a positive constant independent of R, while θ and β are defined by

$$\frac{1}{2} = \theta\left(\frac{1}{2}-\frac{2}{n}\right) + \frac{1-\theta}{q}$$

$$\beta = \frac{n}{q} - \frac{n}{2} > 0$$

and θ satisfies $0 \leq \theta \leq 1$.

Taking squares in (8), using the numerical inequality $A^a B^b \leq C(A+B)^{a+b}$ and observing that $1/q > \theta/2 + (1-\theta)/q$ it follows that

$$C\int u^2 \leq R^{-2\beta} K(R) (J_0(R))^{\theta + 2(1-\theta)/q} \tag{9}$$

where

$$K(R) \equiv R^{2\beta} + \left(\int |u|^q\right)^{\theta(2-q)/q} \tag{10}$$

Notice that the exponents in (10) are positive.

From (7) and (9) we obtain

$$J_4(R) \leq CR^{-2\beta} K(R_0)(J_0(R))^{\theta+2(1-\theta)/q} \tag{11}$$

Since

$$J_0(R) = \frac{1}{4!}\frac{d^4 J_4(R)}{dR^4}$$

(11) is a fourth order differential inequality. Since the integrands of J_s are non-negative it follows from Hölder's inequality that

$$J_1 \leq (J_4)^{1/4}(J_0)^{3/4}$$

This and (11) imply that

$$J_1(R) \leq CR^{-\beta/2}K(R_0)^{1/4}J_0(R)^{\mu} \tag{12}$$

where

$$\mu = \frac{3}{4} + \frac{1}{4}\left(\theta + \frac{2}{q}(1-\theta)\right) > 1$$

The relation (12) is a first order differential inequality in the variable R, since $J_0(R) = J_1'(R)$.

6. Study of the ordinary differential inequality

By explicit integration it is obtained the following lemma.

Lemma. *Let* $0 < \lambda < 1$, $s > 0$, $A > 0$, $y \in C^1$, $y(x) \geq 0$ *and*

$$x^s y(x)^{\lambda} \leq Ay'(x) \quad \text{if} \quad 0 \leq x \leq x_0 \tag{13}$$

If $x_1 > 0$ *then* $y(x) = 0$ *for* $0 \leq x \leq x_1$, *where* x_1 *is defined by*

$$x_1^{s+1} = x_0^{s+1} - \frac{s+1}{1-\lambda}A\left(\frac{Ay'(x_0)}{x_0^s}\right)^{(1-\lambda)/\lambda} \tag{14}$$

Remark: The function $y'(x)$ is going to correspond to the usual energy $J_0(R)$, while $y(x)$ corresponds to the weighted energy $J_1(R)$. That is why we present the formula (14), which involves y' rather than y.

7. Conclusions on the existence of dead core

We apply (13)-(14) to (12) with

$$\lambda = 1/\mu \quad s = \beta\lambda/2 \quad A = CK(R_0)^{\lambda/4}$$

In this way we obtain that

$$u \equiv 0 \text{ in } B_R \text{ if } R < R_1$$

where R_1 is defined by

$$R_1^{s+1} = R_0^{s+1} - CK(R_0)^{\lambda/4}\left(\frac{K(R_0)^{\lambda/4}J_0(R_0)}{R_0^s}\right)^{(1-\lambda)/\lambda} \tag{15}$$

We can assure that there exists dead core (i.e. that $R_1 > 0$) in the following two situations.

1. **Dead core for small energy.** $R_1 > 0$ if $J_0(R_0)$ is small enough.
2. **Dead core for large domains.** If we keep the energy bounded then

$$R_1^{s+1} \quad \text{behaves as} \quad R_0^{s+1} - CR_0^s$$

and hence $R_1 > 0$ for R_0 large enough.

References

1. S.N. Antoncev, On the localization of solutions of nonlinear degenerate elliptic and parabolic equations, Soviet Math. Dokl. 24 (1981), 420-424.
2. F. Bernis, Compactness of the support for some nonlinear elliptic problems of arbitrary order in dimension n, Comm. Partial Differential Equations 9 (1984), 271-312.
3. F. Bernis, Extinction of the solutions of some quasilinear elliptic problems of arbitrary order, Proc. Symp. Pure Math. 45 (1986), Part 1, 125-132.
4. F. Bernis, Existence results for doubly nonlinear higher order parabolic equations on unbounded domains, Math. Ann. 279 (1988), 373-394.
5. F. Bernis, Qualitative properties for some nonlinear higher order degenerate parabolic equations, Houston J. Math. 14 (1988), 319-352.
6. F. Bernis, Finite speed of propagation and asymptotic rates for some nonlinear higher order parabolic equations with absorption, Proc. Royal Soc. Edinburgh 104A (1986), 1-19.
7. F. Bernis, Change of sign of the solutions to some parabolic problems. In: Nonlinear Analysis and Applications, ed. by V. Lakshmikantham, Marcel Dekker, New York, 1987, pp. 75-82.
8. F. Bernis, Elliptic and parabolic semilinear problems without conditions at infinity. Arch. Rational Mech. Anal. 106 (1989), 217-241.
9. F. Bernis, Dead core for higher order elliptic and parabolic problems, to appear.
10. J.I. Diaz, Nonlinear Partial Differential Equations and Free Boundaries, Vol. I, Elliptic Equations. Pitman (Research Notes in Mathematics 106), London, 1985.
11. J.I. Diaz and L. Veron, Local vanishing properties of solutions of elliptic and parabolic quasilinear equations, Trans. Amer. Math. Soc. 290 (1985), 787-814.
12. E. Gagliardo, Ulteriori propieta di alcune classi di funzioni in piu variabili, Ricerche Mat. 8 (1959), 24-51.
13. G.H. Hardy, J.E. Littlewood and G. Polya, Inequalities. Cambridge University Press, Cambridge, second ed., 1952.
14. L. Nirenberg, On elliptic partial differential equations, Ann. Scuola Norm. Sup. Pisa 13 (1959), 115-162.
15. L. Nirenberg, An extended interpolation inequality, Ann. Scuola Norm. Sup. Pisa 20 (1966), 733-737.

THE ANALYSIS OF DIFFUSION CONTROLLED REACTIONS WITH NON-EQUAL DIFFUSIVITIES OF THE REACTANTS

Amable LIÑAN
Escuela Técnica Superior de Ingenieros Aeronáuticos
Universidad Politécnica de Madrid

Abstract

We show, using as a model problem the description of the reaction of a fuel pocket with the oxygen of the environment, how to derive a system of conservation equations to describe the evolution of the temperature and concentrations of the reactants, when these are not initially mixed, or only partially mixed, and the characteristic reaction time is very short compared with the characteristic diffusion time.

1. Introduction and formulation

In the analysis of reactive systems, in particular of combustion systems, we find that the characteristic reaction time, t_c, is often very short when compared with the diffusion time, t_d, or with the residence time in the chamber. Then, the reactants (the fuel and oxygen in combustion systems), if fed independently into the reaction chamber, can only coexist, and with small concentrations, in thin layers (the diffusion flames of combustion systems), where the chemical reaction takes place when the reactants arrive by diffusion from opposite sides of the layer. The products and heat liberated in the layer by the reaction diffuse out in opposite direction to the reactants; their peak concentrations and the peak value of the temperature, reached at the thin layer, is determined by the relative values of the thermal and mass diffusivities of the reactants and products.

The reaction layer becomes infinitely thin in what we call the Burke-Schumann

S. N. Antontsev et al. (eds.), Energy Methods in Continuum Mechanics, 38–48.

limit of infinite chemical reaction rate; namely, in the limit of infinite Damköhler numbers, $t_d/t_c \to \infty$. The position of the reaction surface is determined by the requirement that the reactants must reach the surface in stoichiometric proportions and have zero concentration there. The problem of determining the position of the flame surface is a free-boundary problem, which becomes extremely difficult when, as desirable in combustion systems, the flow is turbulent; then,the velocity fluctuations give a strongly wrinkled fractal character to the flame surface.

Burke and Schumann (1928) showed how the problem of determining the flame surface, and the temperature and concentration of the reactants, could be simplified when the diffusivities of the reactants, D_F and D_O, are equal to the thermal diffusivity, D_T. In this case, we find linear combination of the reactant concentrations and temperature, also called Schvab-Zeldovich coupling functions, not changed by the reactions, which diffuse through the chamber as conserved scalars. One normalized form of these conserved scalar is the mixture fraction Z, widely used in the analysis of turbulent combustion systems, defined so as to be zero in the air feed stream and 1 in the fuel feed stream. See Williams (1985).

The purpose of this note is to show how it is possible to generalize the Burke-Schumann technique to deal with cases where the Lewis numbers, $L_F = D_T/D_F$ for the fuel and $L_O = D_T/D_O$ for the oxygen, are different from unity. For earlier presentatios of this generalization see Liñán (1991a) and (1991b) and Liñán and Williams (1993).

In this note, for simplicity in the presentation, we shall leave out the effects of convection, and thus limit ourselves to the analysis of reactive-diffusive systems. We shall illustrate the method by analysing the evolution with time of a pocket of fuel, of size L, in an unbounded air environment.

If the characteristic reaction time t_c is very small compared with the characteristic diffusion time $t_d = L^2/D_T$, we shall find two stages in the evolution of the system. A first stage, for times $t \sim t_c$, where if the reactants are initially mixed the reaction will take place, without effects of diffusion, until complete depletion of either the fuel or the oxygen. Then, in a second much longer stage, the fuel remaining in the core of the fuel pocket will diffuse out through the products to meet the oxygen, coming in the opposite direction. The reaction will occur in a thin reaction layer that will shrink to diseppear in a time of order t_d.

We shall consider an irreversible chemical reaction of the type

$$F + sO_2 \rightarrow (1+s)P_r + (q) \tag{1}$$

where s mass units of oxygen are consumed per unit mass of fuel to generate $(1+s)$ mass units of products and a quantity q of heat. Although it is not essential for the results of the analysis of the second stage, we shall consider that the mass consumption rate of fuel per unit volume and time, W_F, follows the Arrhenius law

$$W_F = \rho B e^{-E/RT} Y_{F0} Y_{O\infty} y_F y_O \tag{2}$$

where y_F and y_O are the mass fractions of the fuel and oxygen, measured with their maximum initial values Y_{F0} and $Y_{O\infty}$. B is a frequency factor, E is the activation energy, R is the gas constant, ρ the density, and T the temperature.

The conservation equations for the mass fractions y_F and y_O and the temperature, if the density, ρ, specific heat, c_p, and diffusivities are assumed to be constant, take the form

$$\frac{\partial}{\partial t} y_F - \frac{D_T}{L_F} \Delta y_F = -\frac{W_F}{\rho Y_{F0}} \tag{3}$$

$$\frac{\partial}{\partial t} y_O - \frac{D_T}{L_O} \Delta y_O = -S \frac{W_F}{\rho Y_{F0}} \tag{4}$$

$$\frac{\partial}{\partial t} \theta - D_T \Delta \theta = \gamma \frac{W_F}{\rho Y_{F0}} \tag{5}$$

Here $\theta = T/T_\infty$, with T_∞ the ambient air temperature. The main parameters are: $S = sY_{F0}/Y_{O\infty}$, the air/fuel mass stoichiometric ratio, and $\gamma = qY_{F0}/c_p T_\infty$, the chemical heat release parameter, together with the Lewis numbers L_F and L_O. The reaction term $W_F/\rho Y_{F0}$ will be written in the form

$$W_F/\rho Y_{F0} = y_F y_O t_0^{-1} e^{\frac{E}{RT_\infty} \frac{\theta - 1}{\theta}} \tag{6}$$

involving the non-dimensional activation energy E/RT_∞ and the chemical time $t_0 = (Y_{O\infty} B)^{-1} e^{E/RT_\infty}$, evaluated at the initial air temperature T_∞.

The system of equations (3)-(5) will be solved, for $x \in R^3$ and $t > 0$, with the initial conditions

$$t = 0 \; : \; y_F = y_{FI}(x/L) \; , \; y_O = y_{OI}(x/L) \; , \; \theta = \theta_I(x/L) \tag{7}$$

where $y_{FI} \leq 1$, $y_{OI} \leq 1$ and θ_I are non-negative continuous functions of the space variable x, involving a scale L, characterizing the size of the fuel pocket; we shall consider y_{FI}, $1 - y_{oI}$ and $\theta_I - 1$ to be of bounded support. See Fig.1 for a sketch of the distributions.

The existence and uniqueness of the solution of the problem (3)-(7) for $t > 0$ is well known; with y_F, y_O and θ functions of the class C_∞.

The system will be written below in non-dimensional form, measuring x with L, and t with the diffusion time $t_d = L^2/D_T$.

From Eq.(3) and (4) we shall first derive the conservation equation

$$\frac{\partial}{\partial t}(Sy_F - y_O) - \Delta(Sy_F/L_F - y_O/L_O) = 0 \tag{8}$$

free from the reaction term. A similar conservation equation, namely,

$$\frac{\partial}{\partial t}(\gamma y_F + \theta) - \Delta(\gamma y_F/L_F + \theta) = 0 \tag{9}$$

can be derived from Eqs (3) and (5).

These equations must be solved, for $t > 0$, together with the non-dimensional form of Eq (3)

$$\frac{\partial}{\partial t} y_F - \frac{1}{L_F} \Delta y_F = -\delta y_F y_O e^{(E/RT_\infty)(\theta - 1)/\theta} \tag{10}$$

and the non-dimensional form (with $L = 1$) of the initial conditions (7). The non-dimensional Damköhler number $\delta = t_d/t_O$ is the ratio of the diffusion and chemical times.

Notice that when the mass diffusivities are equal to the thermal diffusivity ($L_F = L_O = 1$), Eqs (8) and (9) reduce to the same conservation equation

$$\frac{\partial W}{\partial t} - \Delta W = 0$$

for the "conserved" scalars $(Sy_F - y_o)$ and $(\gamma y_F + \theta)$.

We shall present, in the following, the asymptotic form of the solution of the problem (7)-(10) for large values of the Damköhler number, $\delta \gg 1$.

2. Fast reaction stage for $\delta \gg 1$

The solution of the problem (7)-(10) for $\delta \gg 1$ involves a first reactive stage for times of order t_c; or, when measured with t_d, at non-dimensional times t of order $1/\delta$ such that τ, defined by

$$\tau = t\delta \quad , \tag{11}$$

is of order unity.

When the system (8)-(10) is written in terms of the time variable τ, appropriate to the first stage, the diffusion terms appear multiplied by the factor $1/\delta$. These terms drop out of the equations when we take the formal limit $\delta \to \infty$, with the space and time derivatives assumed to be of order unity. In this limit the equations (8)-(10) simplify to

$$\frac{\partial}{\partial \tau}(Sy_F - y_O) = 0 \tag{12}$$

$$\frac{\partial}{\partial \tau}(\gamma y_F + \theta) = 0 \tag{13}$$

$$\frac{\partial}{\partial \tau}(y_F) = -y_F y_O \; exp\{(E/RT_0)(\theta - 1)/\theta\} \tag{14}$$

for $y_F(x, \tau)$ and the conserved scalars

$$Y = Sy_F - y_O \qquad \text{and} \qquad H = \gamma y_F + \theta \tag{15}$$

which, according to Eqs. (12) and (13), do not vary with τ, and are given by the initial conditions

$$Y = Sy_F - y_O = Y_I(x) = Sy_{FI} - y_{OI} \tag{16a}$$

$$H = \gamma y_F + \theta = H_I(x) = \gamma y_{FI} + \theta \tag{16b}$$

These equations can be used with Eq (14) to calculate the evolution of y_F with τ by means of a quadrature.

For values of τ moderately large compared with unity, but not of order δ, so that the system of equations (12)-(14) remains a valid representation of the system (8)-(10), y_F, y_O and θ will reach the intermediate asymptotic form y_{Fi}, y_{Oi} and θ_i defined by Eqs (16) and the chemical equilibrium condition

$$y_F y_O = 0 \quad , \tag{17}$$

corrcoponding to tho aoymptotic form of (14) for largo τ. In tho fuol domain, Ω_{FI}, where $Y_I > 0$, we will run out of oxygen for $\tau \gg 1$; while in the outer complementary domain Ω_{OI}, where $Y < 0$, we run out of fuel. In the level surface $Y_I = 0$, where the reactants were initially in stoichiometric proportions, the oxygen and the fuel are depleted simultaneously.

At the end of the first stage, at non-dimensional times t such that $1/\delta \ll t \ll 1$, $y_F = y_{Fi}$, $y_O = y_{Oi}$, $\theta = \theta_i$ given by

$$y_{Oi} = 0\,, \qquad S y_{Fi} = Y_I(x) \qquad \text{where } Y_I \geq 0 \text{ at } \Omega_{FI} \tag{18a}$$

$$y_{Fi} = 0\,, \qquad y_{Oi} = -Y_I(x) \qquad \text{where } Y_I \leq 0 \text{ at } \Omega_{OI} \tag{18b}$$

The temperature θ_i at the end of the first stage is given by the relation

$$\theta_i - \theta_I = \gamma(y_{FI} - y_{Fi}) \tag{18c}$$

Notice that the functions y_{Fi}, y_{Oi} and θ_i, appearing in the asymptotic form (18) of the solution of Eqs (12)-(14) for $\tau \to \infty$, although of the class C_0, do not belong to C_1, even if the initial functions y_{FI}, y_{OI} and θ_I are of the class C_∞.

The functions y_{Fi}, y_{Oi} and θ_i have discontinuous derivatives normal to the surface $Y_I(x) = 0$, where y_{Fi} and y_{Oi} are simultaneously equal to zero.

3. Diffusion-controlled stage for $\delta \gg 1$

The fuel remaining, with the mass fraction $y_{Fi} > 0$, at Ω_{FI} diffuses out of the fuel pocket to meet the oxygen, which diffuses from Ω_{OI} to Ω_{FI}. They meet at a surface Γ_f, initially at $Y_I(x) = 0$, where the reaction takes place in a diffusion controlled way, with negligible concentration (in fact of order $\delta^{-1/3}$) of the reactants.

If we follow the method proposed by Burke and Schumann (1928) in their analysis of the case $L_F = L_O = 1$, we find that the conservation equations (8) and (9), together with (17), describe the evolution of the concentrations and temperature, for large values of the Damköhler number, during the long diffusive stage when $1/\delta \ll t \sim 1$. Eq.(17) is obtained from Eq(10) when taking the formal limit $\delta \to \infty$; in this limit the functions and the derivatives appearing in the equation are assumed to be of order unity.

The system of equations (8), (9) and (17) are to be solved with the initial conditions

$$y_F = y_{Fi}\,, \qquad y_O = y_{Oi}\,, \qquad \theta = \theta_i \qquad \text{at } t = 0 \tag{19}$$

as given by Eqs (18).

The solution of this limiting problem, for $\delta \to \infty$, will no longer involve functions y_F, y_O and θ belonging to C_∞, but only to C_0. In the limit $\delta \to \infty$ of infinite reaction rates the chemical production term in Eq (10) becomes localized, as a Dirac delta function of variable strength, on the flame surface, Γ_f, to be determined in terms of t.

The Dirac delta function leads to jumps in the derivatives of y_F, y_O and θ normal to the flame surface. However, these jumps are such that the functions

$$\tilde{Y} = Sy_F/L_F - y_O/L_O \qquad \text{and} \qquad \tilde{H} = \gamma y_F/L_F + \theta \tag{20}$$

must belong, except at $t = 0$, to the class of functions C_1, with continuous derivatives. This is because there are no Dirac delta source terms in the conservation equations (8) and (9) and, therefore, there are no jumps in the spatial derivatives of $\hat{Y}$ and $\hat{H}$. See the sketch of the distributions in Fig.2.

For the description of the solution of the problem (8), (9), (17) and (19), during the second stage $t \sim 1$, we shall use as dependent variables Y and H, defined by (15), and $\hat{Y}$ and $\hat{H}$ defined by Eq (20). In terms of these variables Eqs(8) and (9) take the form

$$\frac{\partial}{\partial t}(Y) - \Delta(\tilde{Y}) = 0 \tag{21}$$

$$\frac{\partial}{\partial t}(H) - \Delta(\tilde{H}) = 0 \tag{22}$$

while the condition (17) implies that

$$y_O = 0\,, \quad Sy_F = Y = L_F\tilde{Y} \qquad \text{where} \quad \tilde{Y} \geq 0 \tag{23a}$$

$$y_F = 0\,, \quad -y_O = Y = L_O\tilde{Y} \qquad \text{where} \quad \tilde{Y} \leq 0 \tag{23b}$$

and, similarly,

$$\theta = \tilde{H} - \gamma\tilde{Y}/S, \qquad H = \tilde{H} + (L_F - 1)\gamma\tilde{Y}/S \qquad \text{where} \qquad \tilde{Y} \geq 0 \tag{24a}$$

$$\theta = H = \tilde{H} \qquad \text{where} \qquad \tilde{Y} \leq 0 \tag{24b}$$

Thus we obtain the system of conservation equations (21)-(22), where the pseudo enthalpies Y and H are known continuous, piece-wise linear, functions of $\tilde{Y}$ and $\tilde{H}$ given by the chemical equilibrium relations (23)-(24) derived from (17). The system (21)-(24) must be solved with the initial conditions

$$t = 0 \; : \quad Y = Y_I(x) = Sy_{FI} - y_{OI}\;, H = H_I(x) = \gamma y_{FI} + \theta_I \tag{25}$$

The solution of this system will provide us with the temperature and concentration field and, in addition, with the time evolution of the flame surface Γ_f, given by the level surface

$$\tilde{Y}(x,t) = 0 \tag{26}$$

The flame temperature there will be $\theta_f = \tilde{H}_f$.

Notice that while $\tilde{Y}$ and $\tilde{H}$ belong to C_1, we can only insure that Y and H belong to C_0; their spatial derivatives may have jumps at Γ_f. Notice also that if the Lewis numbers are equal to unity the problem simplifies considerably, because

$Y = \tilde{Y}$ and $H = \tilde{H}$; which, for $t > 0$, become again functions of the class C_∞, even if they were not so initially.

Concluding remarks

In many practical combustion systems the reactants are not initially partially mixed, as in the example considered here, but are fed in the form of independet jets into the reacting chamber. The conservation equations must then include convective terms associated with the flow.

After an ignition transient stage,whose memory will disappear during the stationary operation of the chamber, the reaction will be diffusion controlled; and described by a straight-forward generalization of the procedure outlined above, if the characteristic chemical time, at the flame temperature T_f, is short compared with the diffusion time. An analysis of the internal structure of the flame, where the reaction term is balanced by the diffusion of heat and mass normal to the thin flame, is required to insure, as shown by Liñán (1974), that indeed the reaction is diffusion controlled.

Some direct numerical simulations of turbulent flow fields with diffusion controlled reactions have already been carried out, see Liñán et al. (1994), using the above formulation.

Let us end this note by noticing that, in the combustion literature,the coupling function $Y = Sy_F - y_O$ is replaced by the mixture fraction variable

$$Z = (Sy_F - y_O + 1)/(S + 1) \qquad (27)$$

equal to zero on the air feed stream, and equal to 1 in the fuel feed stream.

The generalized mixture fraction to be used, with Z, when the Lewis numbers are not equal to 1, can be defined as

$$\tilde{Z} = \frac{Sy_F/L_F - y_O/L_O + 1/L_O}{S/L_F + 1/L_O} = \frac{\tilde{S}y_F - y_O + 1}{\tilde{S} + 1} \qquad (28)$$

where $\tilde{S} = SL_O/L_F$.

The conservation equation (21), when written in terms of $\tilde{Z}$, takes the form

$$\frac{\partial Z}{\partial t} - \frac{1}{L_m}\Delta\tilde{Z} = 0 \tag{29}$$

where $L_m = L_O(1+S)/(1+\tilde{S})$ is a mean value of the Lewis number, and Z is a continuous, piece-wise linear, function of $\tilde{Z}$. The flame surface lies, where y_F and y_O are simultaneously equal to zero, at

$$\tilde{Z} = \tilde{Z}_s = 1/(\tilde{S}+1) \tag{30}$$

where $Z = Z_s = 1/(S+1)$. The chemical equilibrium condition, $y_F y_O = 0$, leads to

$$y_F = 0\,, \qquad \text{where} \qquad 1 - y_O = \tilde{Z}/\tilde{Z}_s = Z/Z_s \geq 0 \tag{31a}$$

$$y_O = 0\,, \qquad \text{where} \qquad y_F = (\tilde{Z}-\tilde{Z}_s)/(1-\tilde{Z}_s) = (Z-Z_s)/(1-Z_s) \geq 0 \tag{31b}$$

Thus providing a relation between Z and $\tilde{Z}$.

Acknowledgment. This research was carried out under partial support of the CICYT, under Contract No. PB94-0400, and also by the INTA, under Contract 4070-0036/1995.

References

1. Burke, S.P. and Schumann, T.E.W. (1928) *Diffusion flames.* Ind. Eng. Chem.20, 998-1004.
2. Liñán, A. (1974) *The asymptotic structure of counter-flow diffusion flames for large activation energies.* Astronautica Acta 1, 1007-1031.
3. Liñán, A. (1991a) *The structure of diffusion flames.* In Fluid Dynamical Aspects of Combustion Theory. Ed. by M. Onofri and A. Tesev. Longman Scientific and Technical pp11-29
4. Liñán, A. (1991b) "El Papel de la Mecánica de Fluidos en los Procesos de Combustiøn". Real Academia de Ciencias, Madrid, Spain.
5. Liñán, A. and Williams, F.A. (1993) "Fundamental Aspects of Combustion". Oxford Univ. Press.
6. Liñán, A.; Orlandi, P.; Verzicco, R. and Higuera, F.J. (1994) *Effects of non-unity Lewis numbers in diffusion flames* . Proc. of the Summer Program 1994. Center of Turbulence Research. Stanford University.
7. Williams, F.A. (1985) "Combustion Theory". Benjamin Cumm ings.

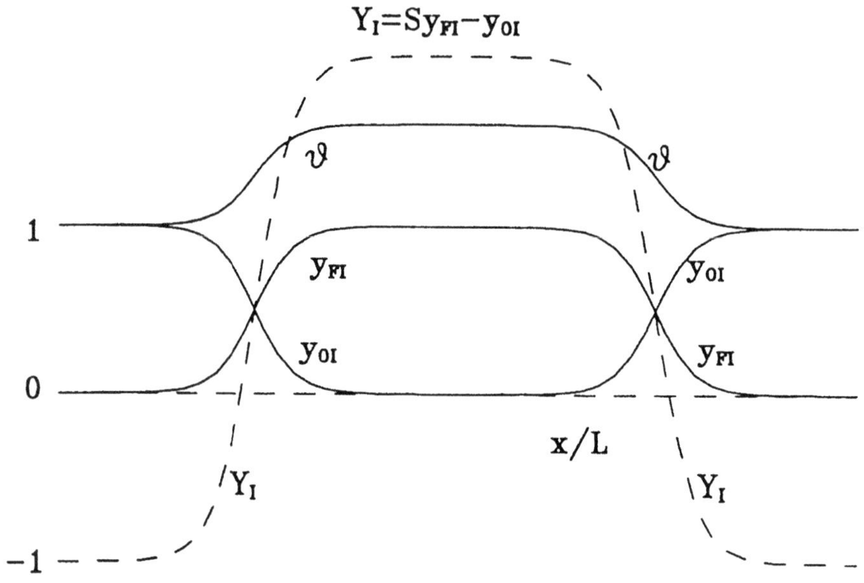

Figure 1. Initial distributions of the mass fractions and temperature for a fuel pocket of size L.

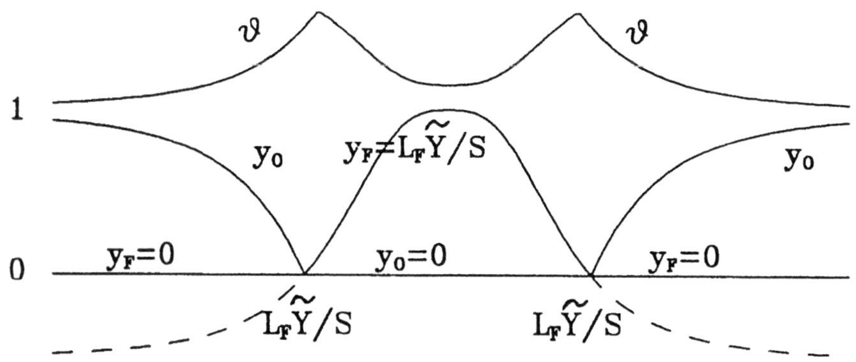

Figure 2. Temperature and mass fraction distribution at a later time during the diffusion controlled stage.

THE BOUNDARY-LAYER PROBLEMS FOR SOME MODELS OF CHANNEL AND FILTRATION FLOWS OF A VISCOUS COMPRESSIBLE FLUID

V.N. Monakhov, E.N. Razinkov, N.V. Khusnutdinova
Lavrentyev Institute of Hydrodynamics
RUSSIA, Novosibirsk, pr.Lavrentyeva, 15

The solvability of initial boundary problems for the systems of equations of combined type describing the flows of a compressible liquid within the framework of the boundary-layer approximations is stated.

1. Boundary layer for the Korteweg model of a viscous compressible liquid.

The Korteweg model is often employed to describe the motion of a viscous compressible liquid using the special state equation (of the type of Van der Waals gas) which allows for the capillary properties of the liquid.

Under the ordinary boundary-layer assumptions the Korteweg equations are transformed into the analog of Prandtl model of a plane stationary boundary layer, whose distinctive feature is its ability to evolve in the two directions: along the wall and along the plane normal to the wall.

According to this, the different types of initial-boundary problems are studied, and their correctness is considered.

The stationary Korteweg equations for a compressible

S. N. Antontsev et al. (eds.), Energy Methods in Continuum Mechanics, 49–65.

liquid are considered in the Felderhopf form [2]:

$$\rho(\vec{u}\cdot\nabla)\vec{u} = \mu\Delta\vec{u} - \rho\nabla(p + \delta\Delta\rho), \ \nabla(\rho\vec{u}) = 0, \ p = p(\rho) \qquad (1.1)$$

where $\vec{u}$ is the flow velocity vector, ρ is the density of the liquid, μ and δ are the positive variables taking account of viscous and capillary properties of the liquid, $p = p\ (\rho)$ is the pressure; $p = p_0/\rho + \int_1^\rho s^{-2}p(s)ds + const.$

Substitute the standard boundary-layer relationships in (1.1) $(u,\rho,p,x,y) = (\bar{u},\bar{\rho},\bar{p},\bar{x},\varepsilon\bar{y})$, $(\mu,\delta) = \varepsilon(\bar{\mu},\bar{\delta})$, assuming additionally the weak vertical mixing in the liquid, $v = \varepsilon\bar{v}$. Discarding the expressions for positive orders of the small parameter $\varepsilon>0$, yields the following boundary-layer equations for the finite values of $\bar{u},\bar{p},..$ (the dash above the symbols will be omitted)

$$\rho(u^2/2 + p + \delta\rho_{yy})_x = \mu u_{yy}, \ (\rho u)_x = 0, \ p = p(\rho). \qquad (1.2)$$

The initial conditions for the sought functions are given in (1.2)

$$\rho|_{x=0} = \rho_0(y), \ u|_{x=0} = u_0(y), \ y \in \Omega = [0,L], \qquad (1.3)$$

which admits the integration of the second equation (1.2):

$$\rho u = R(y), \ R = u_0(y)\rho_0(y).$$

After the appropriate expansion of independent variables and sought functions the problem (1.2) and (1.3) becomes:

$$(m + \rho_{yy})_x = \mu\rho^{-1}(\rho^{-1})_{yy}, \ \rho|_{x=0} = \rho_0(y), \qquad (1.4)$$

where $m = \frac{1}{2}\rho^{-2} + p(\rho)$ and without loss of generality $R(y)=1$ is set, i.e. $u = \rho^{-1}$.

The sought-for function $\rho(x,y)$ in (1.4) satisfies one of the following boundary conditions:

$$(\rho - \rho_1(x), \rho_y)|_{y=0} = 0, \quad x \in [0,T] \qquad (1.5)$$

$$\rho_y|_{y=0,L} = 0, \quad x \in [0,T] \qquad (1.6)$$

$$(\rho - \rho_1(x))|_{y=0} = (\rho - \rho_2(x))|_{y=L} = 0, \quad x \in [0,T] \qquad (1.7)$$

Let us integrate Eq.(1.4) with respect to $\xi = x$ within the limits $[0,x]$, $x \in [0,T]$, multiply the relationship by the function $\varphi \in W_2^1(\Omega)$, $\Omega = (0,L)$ and integrate the result with respect to Ω:

$$\int_\Omega [l_y \varphi_y - (m - m_o + \int_0^x \rho^{-4} \rho_y^2 d\xi)\varphi] dy = 0 \qquad (1.8)$$

where $m_o = m(\rho_o)$, $l = \rho - \rho_o - \frac{1}{2}\int_0^x \rho^{-2}(\xi,y)d\xi$.

Let us specify $q(\rho) = m + \rho_{yy}$ and use the notation of the norms and spaces accepted in [3].

Definition 1. The function $\rho(x,y)$, $\rho \in C(0,T;W_\infty^2(\Omega))$, $|\ln(\rho)| \leq N < \infty$, $q(\rho) \in W_\infty^1(0,T;L_\infty(\Omega))$ satisfying the relationships (1.4) and (1.5), ((1.6) or (1.7)) for almost every $(x,y) \in Q_T$ is called strong solution of the problem (1.4) and (1.5) ((1.6) or (1.7) respectively) in $Q_T = (0,T)\times(0,L)$. **Definition 2.** The function $\rho(x,y)$, $\rho \in C(0,T;W_2^1(\Omega))$, $|\ln(\rho)| \leq N < \infty$, satisfying the integral identity (1.8) for any $x \in [0,T]$, $\varphi \in W_2^1(\Omega)$ is called weak solution of the problem (1.4) and (1.6) in Q_T.

If additionally to (1.8) $\varphi(L) = 0$, $\rho|_{y=0} = \rho_1(x)$ in the problem (1.4) and (1.5), and $\varphi|_{y=0,L} = 0$ $(\rho-\rho_1)|_{y=0} = (\rho-\rho_2)|_{y=L} = 0$ in the problem (1.4) and (1.7), then $\rho(x,y)$ is said to be the weak solution of the problem (1.4) and (1.5) (or (1.7)).

Theorem 1. Let $m(\rho) \in W_\infty^1(a,b)$ $\forall [a,b] \subset (0,\infty)$ and $\rho_o \in W_2^1(\Omega)$, $\rho_1 \in C[0,T]$, $|\ln \rho_o| + |\ln \rho_1| \leq N < \infty$, $\rho_o(0) = \rho_1(0)$. Then there exist a unique weak solution of the problem (1.4) and (1.5) in Q_T, $\forall$ $T \in (0,\infty)$. If additionally $\rho_o \in W_\infty^2(\Omega)$, $(\rho_o)_y|_{y=0} = 0$, then the weak solution $\rho(x,y)$ is also a strong one.

In this case the appropriate norms of weak and strong solutions are independent of T.

The proof of the existence of the strong solution $\rho(x,y)$ of the problem (1.4) and (1.5) when the theorem conditions are satisfied is equivalent to the ascertainment of the solvability of the operator equation

$$\rho = \int_0^y (y-s)A(\rho,\rho_y)(x,s)ds + \rho_1 \equiv \Lambda(\rho,\rho_y) \qquad (1.9)$$

where $A = -m(\rho) + D(\rho)(q_o + \int_0^x [D^{-1}(\rho)f(\rho,\rho_y)](\xi,y)d\xi)$, $m_o = m(\rho_o)$, $D=\exp[-\int_0^x \rho^{-3}(\xi,y)d\xi]$, $f = \rho^{-3}m(\rho) + 2\rho^{-4}v_y^2$, $q_o(y) = m_o + \rho_{oyy}$

By virtue of the assumptions on the smoothness of the preset functions included in (1.4) and (1.5) for a sufficiently small thickness L of a boundary layer nearby the initial profile of the function $\rho|_{y=0} = \rho_1(x)$ the contractibility of Λ operator is stated, and, hence, the solvability of Eq. (1.9). Thus, the validity of the following uniform estimate with respect to T is proved

$$\|\rho\|_{W_2^1(\Omega)} \leqslant C(\|\rho_o\|_{W_2^1(\Omega)} + C_o)$$

To prove the existence of the weak solution for the problem (1.4) and (1.5) the given $\rho_o(y)$ is approximated by the smooth functions $\rho_o^k(y)$ satisfying the existence conditions of the strong solution $\rho^k(x,y)$ of the appropriate approximate problem. Assuming $\Delta\rho = \rho^n - \rho^k$ $\forall(n,k)>0$, $w(x) = \|\Delta\rho\|_{W_2^1(\Omega)}$, $\Delta\rho_o = \rho_o^n - \rho_o^k$, $\Delta m_o = m_o^n - m_o^k$ with Gronuolla lemma [3] as applied to (1.8) for $\varphi = \Delta\rho - \Delta\rho|_{y=L}$ the inequality is stated:

$$w(x) \leqslant C(\|\Delta\rho_o\|_{W_2^1(\Omega)} + \int_\Omega |\Delta m_o| dy)e^{C_0 x} \qquad (1.10)$$

The convergence of $\rho(x,y)$ sufficient for the passage to the limit in integral identity (1.8) follows from (1.10). Inequality (1.10) is apparent, and provides the uniqueness of the weak solution of the problem (1.4) and (1.5).

Suppose that the given functions $m = \rho^{-2} + p(\rho)$, $\rho_o = \rho_o(y)$ possess the following properties:

(i) $m(\rho) \in C[a,b]\ \forall[a,b]\subset(0,\infty)$; $m(\rho_1)< m(\rho_2)$ for $\rho_2<\rho_1$, $m(\infty)=0$, $-p(\rho)\leqslant C_o<\infty$; $\rho_o\in W_2^1(\Omega)$, $|\ln\rho_o|\leqslant N<\infty$;

(ii) $m(\rho)\geqslant as^{-4}$ for $s\geqslant s_o$, $a>0$;

(iii) $F(\alpha)\equiv\int_\Omega(\rho_{oy}^2-\rho_o m_o+\int_\alpha^{\rho_o} m(s)ds)dy\leqslant-\beta<0$ for a certain $\alpha>0$.

Theorem 2. Let the condition (i) be held, as well as one of the conditions (ii) or (iii). Then there is a unique solution of the problem (1.4) and (1.6) in Q_T $\forall$ $T>0$.

With additional assumptions $\rho_o\in W_\infty^2(\Omega)$, $\rho_{oy}|_{y=0,L}=0$ the solution $\rho(x,y)$ is a strong one, and it also $m(\rho)\in W_\infty^1(a,b)$ $\forall[a,b]\in(0,\infty)$, then $\rho(x,y)$ is a classical solution $(\rho,\nabla\rho,\rho_{yy},\rho_{yyx})\in L_\infty(Q_T)$.

The appropriate solution norms are independent of T, if (ii) is valid.

Theorem 3. Let the conditions (i) and $\rho_i\in W_1^1(0,T)$, $|\ln\rho_i|\leqslant$ $\leqslant N<\infty$, $i=1,2$ be satisfied.

Then there is a unique weak solution of the problem (1.4) and (1.7) in Q_T $\forall T>0$.

With additional suggestion $\rho_o\in W_\infty^2(\Omega)$ this solution is a strong one, and if also $m(\rho)\in W_\infty^1(a,b)$ $\forall[a,b]\subset(0,\infty)$, $\rho_i\in W_\infty^1(0,T)$, $i=1,2$, then $\rho(x.y)$ belongs to the class $(\rho,\nabla\rho,\rho_{yy},\rho_{yyx})\in L_\infty(Q_T)$.

The appropriate solution norms are independent of T, if $[\rho_{1x},\rho_{2x},(\rho_1-\rho_2)^2]\in L_1(0.\infty)$.

Proof. The proof of the theorems 2 and 3 like that of the theorem 1, begins from the construction of the smooth solution $\rho(x,y)$, $(\rho,\nabla\rho,\rho_{yy},\rho_{yyx})\in L_\infty(Q_T)$ in the application (small with respect to T) of the Shauder principle of immovable point to the equivalent operator equation of (1.9) type.

Consider the problem (1.4) and (1.6) in more detail. For the purpose of extension of local solution of this problem the a priori estimates of $\rho(x,y)$ are established over an arbitrary finite interval (0,T), uniform with respect to T.

Let us integrate Eq. (1.4) and the result of its multiplication by $\rho(x,y)$ with respect to Q_t as follows:

$$\int_\Omega m(\rho)|_{x=t}dy + \iint_{Q_t}\rho^{-4}\rho_y^2 dxdy = \int_\Omega m_o dy$$

$$\int_\Omega (\rho_y^2 + v(\rho,\alpha) - \rho m)|_{x=t}dy = F(\alpha),\ v = \int_\alpha^\rho m(s)ds,$$

where $F(\alpha)$ is defined in (iii).

The uniform estimates $\|\rho^{-1},\rho_y\|_{L_2(\Omega)} \leqslant C$ with respect to T follow directly from the relationships obtained, whence

$$\int_\Omega |(ln\rho)_y|dy \leqslant \|\rho^{-1}\|_{L_2(\Omega)}\|\rho_y\|_{L_2(\Omega)} \leqslant C^2$$

that in its own turn results in the estimate

$$\sup(\rho^{-1}, |m|) = C_o < \infty$$

Integrating the equation (1.4) with respect to $q = m + \rho_{yy}$ yields the representation

$$q = D(\rho)\{q_o + \int_0^x D^{-1}(\rho)f(\rho,\rho_y)(s,y)ds\}$$

in which D and f are determined in (1.9), which makes it possible to estimate $\|q\|_{L_1(\Omega)}$, $\|q\|_{L_\infty(\Omega)}$ and thus $\|\rho_{yy}\|_{L_\infty(\Omega)}$.

The estimates obtained are sufficient to extend the smooth solution of the problem (1.4) and (1.6) to $(0,T)$, $\forall$ $T>0$.

The required a priori estimates are stated similarly in the problem (1.4) and (1.7).

To prove the existence of a strong solution of the problems (1.4) and (1.6), (1.4) and (1.7) the functions $m(\rho)$, $\rho_0(y)$, $\rho_1(x)$, $\rho_2(x)$ are approximated by smooth functions m^k, ρ_0^k, ρ_1^k, ρ_2^k and the estimates of the approximate solution norms required for passage to the limit are established.

Then the passage to the limit of the weak solutions of these problems is accomplished in the same way as in the proof of the Theorem 1 and their uniqueness is stated.

2. Conjugation of channel and filtration flows of a viscous liquid.

The problem of conjunction of the flows of a viscous liquid in the holes or open river beds (channels) with its filtration flows in the ambient viscous medium is studied. Owing to considerably high relativity of the velocity of conjugate flows they can interact only via an intermediate boundary layer nearby an interface.

Different variants of conjugation of such flows are proposed within the framework of the boundary-layer approximations for both flows, the solvability of boundary problems for mutually perpendicular boundary layers in the hole and adjacent porous medium is proved, the class of self-similar solutions is constructed.

Let us choose the direction of the OX axis parallel to the liquid flow in the hole, while the filtration flow is directed along the OY axis. With the ordinary boundary-layer assumptions on predominant movement of liquid flows, the Navier-Stokes equations describing them will be substituted by the following ones:

$$\vec{u}\nabla u = \mu u_{yy} - p_x, \quad \nabla\vec{u} = 0, \quad (x,y) \in D_1; \tag{2.1}$$

$$\vec{u}\nabla u = \mu v_{xx} - p_y - \lambda v, \quad \nabla\vec{u} = 0, \quad (x,y) \in D_2 \tag{2.2}$$

where $\vec{u} = (u,v)$ is the vector of liquid flow velocity with the density $\rho=1$, μ=const is the viscosity, $p = p_o + \rho g x$, p_o is the pressure, $\vec{g} = (-g,0)$ is the gravity acceleration vector. The projections of external force are included in (2.2). According to the filtration theory assumptions the force is represented in the form $\vec{F} = -\lambda\vec{u}$, $\lambda = m\mu k^{-1}$, m is the porosity, k is the permeability of porous medium.

The flow velocity vector $\vec{u}$ and pressure p are assumed constant at the line $\Gamma = \overline{\partial D_1} \cap \overline{\partial D_2}$.

Note that with regard for the direction of filtration flow after the change

$$x = \eta,\ y = -\xi,\ u = V,\ v = -U$$

of the equation (2.2) for $\lambda=0$ we shall transform into the equations (2.1) for $U(\xi,\eta)$, $V(\xi,\eta)$.

Problem 1 of the conjugation of filtration and free flows at the hole wall.

Let $D_1=\{x>0,\ 0<y<h\}$ be the zone corresponding to the hole symmetrical about $y=0$, while $D_2=\{x>0,\ y<0\}$ is the zone of liquid filtration. The liquid flows in D_1 and D_2 may be described by the solution of the following boundary problems

$$u|_{y=h} = (u_y, v)|_{y=0} = 0,\ x \geqslant 0;\quad u|_{x=0} = u_o(y),\ y \geqslant 0; \tag{2.3}$$

$$(\vec{u} - \vec{u}_o)|_{x=0} = 0,\ y \geqslant 0;\quad v|_{y=0} = v_1(x),\ x \leqslant 0; \tag{2.4}$$

for the equations (2.1) and (2.2) respectively, where $\vec{u}_0=(u_o(y), v(+0,y))$ and $u_o(y)$, $v_1(x)$ are the arbitrarily given functions.

Problem 2 of the conjugation of filtration and free flows at the hole entry.

In this problem two mutually perpendicular boundary layers near the hole wall $\{y=0,\ 0<x<X\}$ and the line $\{x=0, -H<y<h\}$ which consists of an impenetrable boundary of porous medium $\{x=0, -H<y<h\}$ and the hole entry $\{x=0, 0<y<h\}$ are joined.

According to this in the domains $D_1=\{0<x<X, 0<y<h\}$ and $D_2=\{x<0, -H<y<h\}$ the following problems on extension of boundary layers (2.1) and (2.2) arise

$$\vec{u}|_{y=0} = 0;\quad u|_{y=h} = u_1(x);\quad u_x|_{x=0} = 0,\quad x>0; \tag{2.5}$$

$$(u - u_o, v)|_{x=0} = 0; \quad v|_{y=-H} = v_1(x), \quad x<0; \tag{2.6}$$

$$\lim_{x\to-\infty} v(x,y) = v_\infty(y), \quad -H < y < h; \tag{2.7}$$

$$v_\infty v'_\infty + p_y + \lambda v_\infty = 0, \quad v_\infty(-H) = v_1(\infty), \quad y \geqslant -H \tag{2.8}$$

Here $u_o(y)=0,\ y<0$, and

$$u_o(y) = \frac{1}{2\mu}p_x(0)y(y-h) + u_1(0)\frac{y}{h}, \quad y>0$$

is determined from the boundary value problem solution

$$\mu_o u''_o - p_x(0) = 0, \quad u_o(0) = 0, \quad u_o(h) = u_1(0),$$

which is a consequence of (2.1) and (2.5).

The functions $u_1(x), v_1(x)$ in (2.5) and (2.6), $p = p_1(x)$, $x\geqslant 0$ and $p=p_2(y)$, $y\leqslant 0$ $(p=p_1(0),\ y>0)$ in (2.1) and (2.2), as well as the velocity $v_\infty(y)$ of external flow in the porous medium preset at ∂D, $D = D_1 \cup D_2$ are subjected to the ordinary suggestions of the boundary-layer theory

$$(p_1, p_2, u_1, u_2) \in C^{2+\alpha}(\partial D),\ \alpha>0;\ p'_1<0,\ x\geqslant 0;\ p'_2<0,\ y\leqslant 0; \tag{A}$$
$$v_\infty>0,\ y \geqslant -H;\ (u_1, u'_1)>0,\ x>0,\ u_1(0)>0,\ u'_1(0)=0;$$
$$(v_1, v'_1)>0,\ x<0,\ v_1(0)=0,\ v'_1(0)>0;\ y=-H,\ x\to 0,$$
$$\mu v''_1 - p'_2(-H) - \lambda v_1 = 0(x^2)$$

Theorem. Let the suggestions (A) hold. Then in the domain D $\forall(x,H) > 0$ for a certain h there is a solution $u(x,y)$, $v(x,y)$ of the problem 2 (2.10, (2.2), (2.5)-(2.8) with the following properties

$$(u, u_y, u_{yy}) \in C(D_1),\ (v, u_x, v_y) \in C(\Omega_1),\ \forall\ \bar{\Omega}_1 \subset D_1; \tag{a}$$

$$(v, v_x, v_{xx}) \in C(D_2),\ (u, u_x, v_y) \in C(\Omega_2),\ \forall\ \bar{\Omega}_2 \subset D_2;$$

$$u(x,y)>0,\ u_y|_{y=0} \geqslant m_1 > 0,\ (x,y) \in D_1; \tag{b}$$

$$v(x,y)>0,\ v_x|_{x=0} \geqslant m_2 > 0,\ (x,y) \in D_2$$

The problem 2 decomposes into three successively solved problems on extension of boundary layers in the domains D_1, $D_3=\{x<0,-H<y<0\}$, $(D_2 \backslash D_3)$ for each the stated existence problem being proved analogously to the classical Prandtl equations of boundary layer.

To construct the self-similar solutions of the equations (2.1) and (2.2) let us introduce the stream function $\psi(x,y)$ assuming $u=\psi_y$, $v=-\psi_x$. Then the equations (2.1) and (2.2) admit the self-similar solutions of the type

$$\psi = y^k\varphi(\xi),\ \xi = y(nx + x_o)^{-\frac{1}{n}},\ x_0 = const,$$

where the constants k and n are connected by the relationships $k=n-1$ for Eq. (2.1) and $k=1-n$ for (2.2).

The conditions on the preset functions $p(x,y),\lambda(x,y)$ involved in the equations arise naturally:

$p_x = \delta_1(nx + x_o)^{\frac{k-3}{n}}$ in (2.1); $p_y = \delta_2 y^{k-3n}$, $\lambda = -\delta_3 y^{-2n}$ in (2.2) σ_i=const, if they hold, (2.1) and (2.2) are transformed in the following equations for the function

$$L^{(m)}\varphi \equiv \sum_{i=0}^{3} \lambda_i^{(m)}\frac{d^i\varphi}{d\xi^i} - f^{(m)} = 0,\ m = 1,2 \qquad (2.9)$$

Here $f^{(m)} = f^{(m)}(\xi)$, $\lambda_i^{(m)} = \lambda_i^{(m)}(\xi,\varphi,\varphi')$ are expressed in terms of the coefficients (2.1) and (2.2). Consider the particular case

$$k = 3,\ n = 4,\ \varphi(\xi) = c \quad \text{when } m = 1 \text{ in } (2.9)$$

corresponding to the flow of the Poiseuille type in $D_1=\{x>0\}$

$$u = cy^2,\ v = 0,\ p_x = 2c\mu,\ c = const.$$

The joining conditions and setting of the external filtration flow result in the boundary value problem for (2.9) when m=2 ($k=3,n=-2$)

$$\varphi(0) = \frac{c}{3},\ \varphi'(0) = 0,\ \xi^{-1}\varphi|_{\xi=\infty} = c_o.$$

3. The model of a filtration boundary layer with stagnation zones.

As is known [5], with $p'(x)>0$ there is no smooth solution to the problem on continuation of the boundary layer

(2.1) subject to the condition $u(x,y)>0$ with $(x,y)\in\{0<x\leqslant X, y\geqslant 0\}$ due to separation of the boundary layer. Consequently, the conjugation of the two smooth mutually perpendicular boundary layers with positive pressure gradient is also possible only over a small interval. The statement of the problem on continuation of the boundary layer involving the separation line was proposed in [7].

Here we shall consider the analogous formulation of the problem for the filtration boundary layer which is described by the analog of Eqs. (2.1):

$$\vec{u}\nabla u = \mu u_{yy} - \lambda u - \varphi(x/u^2), \quad \nabla\vec{u} = 0$$

where the notation of the previous section is retained, while the function simulates the pressure in the outer flow [7]

$$\varphi(x/u^2) = p'(x), \text{ when } u > 0, \ \varphi(x/0) = 0$$

Passing to Mises variables, we obtain the following problem: the function $w(x,\psi) = u^2$ satisfying the equation

$$w_x = \mu\sqrt{w}\,w_{\psi\psi} - 2\lambda\sqrt{w} - 2\varphi(x/w) \qquad (3.1)$$

and the boundary condition

$$w(x,0) = 0, \ w(0,\psi) = w_0(\psi), \ \psi \geqslant 0 \qquad (3.2)$$

in the domain $D = \{(x,\psi) \in (0,X]\times(0,\infty)\}$ should be found.

Definition. The continuous nonnegative function ω bounded in $\bar{D}$ is called the generalized solution to the problem (3.1) and (3.2) in the domain D, if the conditions (3.2) and the following conditions hold:

a) there is a generalized derivative ω bounded in $\bar{D}$;

b) the generalized derivative $\frac{\partial}{\partial\psi} w^{3/4}$ is square integrable in any finite subdomain D;

c) for any function $f(x,\psi) \in \overset{0}{C}{}^1(D)$ the following identity is valid

$$\iint_D [wf_x - \mu\sqrt{\bar{w}}\, w_\psi f_\psi - (\mu w_\psi^2/2\sqrt{\bar{w}} + 2\lambda\sqrt{\bar{w}} + 2\varphi(x/w))f]dD +$$
$$+ \int_0^\infty w_0(\psi)f(0,\psi)d\psi = 0 \qquad (3.3)$$

Let us assume that

(i) $w_0(\psi)>0$ as $\psi>0$, $w_0(0)=0$, $\lim_{\psi\to\infty} w_0(\psi)=w_\infty=const>0$

(ii) $w_0(\psi)\in C^1[0,\infty)$, $p'(x)\geqslant 0$, $p(x)\in C^{2+\alpha}[0,X]$, $\alpha\in(0,1)$

Theorem 3.1. The generalized solution to the boundary value problem (3.1) and (3.2) in the domain D exists $\forall\ X > 0$.

Let construct the solution $w(x,\psi)$ to the boundary problem (3.1) and (3.2) as a limit $\varepsilon\to 0$ of smooth positive solutions $w(x,\psi,\varepsilon) \equiv w_\varepsilon$ of boundary problems

$$w_x = \mu\sqrt{\eta}w_{\psi\psi} + \chi(\eta)(w - \varepsilon), \quad (x,\psi) \in D_\varepsilon \qquad (3.4)$$

$$w(0,\psi,\varepsilon)=w_{0\varepsilon}(\psi),\ w(x,0,\varepsilon)=\varepsilon,\ w(x,1/\varepsilon,\varepsilon)=w_{0\varepsilon}(1/\varepsilon) \qquad (3.5)$$

Here $D_\varepsilon = \{(x,\psi) \in (0,X]\times(0,1/\varepsilon)\}$, $\eta = \eta(\varepsilon,\xi) \in C^3(-\infty,\infty)$ is the cuttoff function such that $\eta(\varepsilon,\xi) = \varepsilon/2$ when $\xi \leqslant \varepsilon/2$, $\eta = 2M$ when $\xi \geqslant 2M$, $\eta = \xi$ when $\varepsilon \leqslant \xi\leqslant M$

$$\chi(\eta) = -\frac{2p'(x)}{\eta + \varepsilon} - 2\lambda\frac{\sqrt{\eta}}{\eta + \varepsilon}$$

$w_{0\varepsilon}(\psi)$ are the triply differentiated functions satisfying the conditions of concordance of the boundary data at the points $(0,0)$, $(0,1/\varepsilon)$ up to the 2nd order inclusively and uniformly converging as $\varepsilon\to 0$ to the function $w_0(\psi)$, with

(iii) $\varepsilon \leqslant w_{0\varepsilon} \leqslant M_1$, $|\frac{dw_{0\varepsilon}}{dx}| \leqslant M_1$, $M_1 > (\sup_{\psi\geqslant 0} w_0\ ,\ \sup_{\psi\geqslant 0} w_0')$

Since the parabolic equations (3.4) do not degenerate and have Holder bounded coefficient, there are smooth solutions $w(x,\psi,\varepsilon)$ to the boundary problems (3.4) and (3.5).

Using the generalized principle of maximum the estimates are proved like in [5] and [7] with $(x,\psi) \in D_\varepsilon$

$$\varepsilon \leqslant w(x,\psi,\varepsilon) \leqslant M_2 \qquad (3.6)$$

$$|\frac{\partial w}{\partial\psi}| \leqslant M_2 \qquad (3.7)$$

$$|w_1(x,\psi,\varepsilon) - w_2(x,\psi,\varepsilon)| \leqslant |x_1 - x_2|^{1/2} \tag{3.8}$$

In this case the constant M_2 is independent of ε, which makes it possible to set in (3.4) $\eta(\varepsilon,w) \equiv w(x,\psi,\varepsilon)$:

$$w_x = \mu\sqrt{w}\,w_{\psi\psi} + \varphi_0(x,w)$$

where $\varphi_0(x,w) = -2[p'(x) + \lambda\sqrt{w}]\,\dfrac{w-\varepsilon}{w+\varepsilon}$

In view of the estimates (3.6)-(3.8) the sequence w_ε is compact in $C^\alpha(D_m)$, $\forall\ m > 0$, $0 < \alpha < 1/2$, and, consequently, one can separate a converging subsequence $w_{\varepsilon k}$ from it, (let us denote it as w_ε).

It is obvious, that for w_ε the following identities hold

$$\iint_D [wf_x - \mu\sqrt{w}\,w_\psi f_\psi - (\mu w_\psi^2/\sqrt{w} - \varphi_0(x/w))f]dD_\varepsilon + \\ + \int_0^{1/\varepsilon} w_0(\psi)f(0,\psi)d\psi = 0 \tag{3.9}$$

for any function $f(x,\psi) \in \overset{0}{C}{}^1(D_\varepsilon)$.

To justify the passage to the limit in the identity (3.9) for $w(x,\psi,\varepsilon)$ as $\varepsilon \to 0$ let us obtain the additional integral estimates.

We set $\iint_{D_m} f\cdot F\; dD = (f,F)$. The following inequality is true

$$(w^{-\frac{1}{2}-\delta}, w_\psi^2) \leqslant M_3,\quad 0 \leqslant \delta < 1/4 \tag{3.10}$$

where the constant M_3 does not depend on ε.

The estimate (3.10) is proved on the basis of the following identities

$$\mu(w^{-1/2}, w_\psi^2) = -2\int_0^{1/m} [w(x,\psi,\varepsilon) - w(0,\psi,\varepsilon)]d\psi + \\ 2\int_0^X \sqrt{w}w_\psi dx\Big|_0^{1/m} + 2(\varphi_0,1) \tag{3.11}$$

$$\mu(w^{\frac{1}{2}}, w_{\psi\psi}^2) = -1/2\int_0^{1/m} w_\psi^2 d\psi\Big|_0^X - ((\varphi_0, w_\psi)'_\psi, 1) + (\varphi'_{0w}, w_\psi^2) \tag{3.12}$$

$$(\varphi_0, w^{-\delta}) = \frac{1}{1-\delta}\int_0^{1/m} w^{1-\delta} d\psi \Big|_0^X - \mu(w^{\frac{1}{2}-\delta}, w_{\psi\psi}),\ 0 \leqslant \delta < \frac{1}{4} \quad (3.13)$$

Since all summands of the right-hand part (3.11) are modulus-bounded, the value $(w^{-1/2}, w_\psi^2)$ is also bounded, i.e. (3.10) is valid with $\delta=0$.

The boundedness $(w^{1/2}, w_{\psi\psi}^2)$ uniform with respect to ε follows from (3.12). Let us use the inequality

$$(w^{\frac{1}{2}-\delta}, w_{\psi\psi}) \leqslant \frac{1}{2}(w^{1/2}, w_{\psi\psi}^2) + \frac{1}{2}(w^{\frac{1}{2}-2\delta}, 1)$$

and estimate $(\varphi_0, w^{-\delta})$, which is possible as $\delta \in (0, 1/4)$.

Finally, if we transform $(w^{\frac{1}{2}-\delta}, w_{\psi\psi})$ in (3.13) using the integration by parts, then the required estimate (3.10) follows from (3.13).

Let us consider the identity (3.9) for $w(x,\psi,\varepsilon)$ and turn to the limit as $\varepsilon\to 0$ $\forall f \in C^1(D)$ term by term in it, beginning from, e.g. the integral $(w^{-\frac{1}{2}} w_\psi^2, f)$.

Let us specify $F(x,\psi) = w^\delta(x,\psi)$, $\delta \in (0,1/4)$, $\Phi(x,\psi) = w^{-\frac{1}{2}-\delta} w_\psi^2$. Denote the same functions with $w = w_\varepsilon$, $w_\psi = w_{\varepsilon\psi}$ by F_ε, Φ_ε and construct the equality for the arbitrary function $f(x,\psi) \in \overset{0}{C}{}^1(D_m)$

$$(F_\varepsilon, \Phi_\varepsilon f) - (F, \Phi f) = (F_\varepsilon - F, \Phi f) + (F_\varepsilon, (\Phi_\varepsilon - \Phi) f)_{\Omega_0} + \\ + (F_\varepsilon, (\Phi - \Phi_\varepsilon))_{\Omega_1} \quad (3.14)$$

where $\Omega_0 = \{(x,\psi) \in D_m,\ |F_\varepsilon| \leqslant \varepsilon_0\}$, $\Omega_1 = D_m \backslash \Omega_0$, $0 < \varepsilon \leqslant \varepsilon_0$.

The first term in the right-hand part (3.14) tends to zero as $\varepsilon\to 0$ due to the uniform convergence of F_ε to F in D_m for all $m>0$. The integral with respect to Ω_0 with small ε_0 is uniformly small with respect to ε. With fixed ε_0 the integral with respect to the domain Ω_1 tends to zero, since in this domain the derivatives $w_{\varepsilon\psi}$ satisfy the Holder condition and, consequently, $\|w_{\varepsilon\psi} - w_{\psi}\|_{C(\Omega_1)} \to 0$. Since ε_0 is arbitrary, $\lim\limits_{\varepsilon\to 0} (F_\varepsilon, \Phi_\varepsilon f) = (F, \Phi f)$, $\forall\ f \in \overset{0}{C}{}^1(D_N)$. The passage to the limit as $\varepsilon\to 0$ in the other terms is performed in the same way. As a result we obtain the identity (3.3) for $w(x,\psi)$.

Thus, the limiting function $w(x,\psi)$ possesses a generalized derivative w_ψ which is bounded by virtue of (3.7). In view of (3.10), the generalized derivative $\frac{\partial}{\partial\psi} w^{\frac{3}{4}}$ is square integrable and satisfies the identity (3.3) for any function $f(x,\psi) \in \overset{0}{C}{}^1(D)$, i.e. is the generalized solution to the boundary-value problem (3.1)–(3.2).

Let us emphasize some of the properties of the generalized solution.

Theorem 3.2. Let $w(x,\psi)$ is the generalized solution to the boundary value problem (3.1)–(3.2).

1. If $w_0'(\psi) \geqslant 0$ with $\psi \geqslant 0$, $p'(x) \geqslant 0$, $x \in [0,X]$, then

$$\frac{\partial w}{\partial \psi} \geqslant 0, \ (x,\psi) \in D \tag{3.15}$$

2. If $\mu\sqrt{w_0(\psi)}w_0'' - 2\lambda\sqrt{w_0} - 2p'(0) < 0$, $p''(x) \geqslant 0$, $x \in [0,X]$, then

$$\frac{\partial w}{\partial x} \leqslant 0, \ (x,\psi) \in D$$

Since as $\varepsilon\to 0$ the generalized solution to the boundary problem (3.1), (3.2) is constructed as a limit of smooth positive solutions $w(x,\varepsilon,\psi)$ of Eq. (3.1), then it suffices to establish the properties 1 and 2 for the functions $w(x,\psi,\varepsilon)$, $(x,\psi) \in D_\varepsilon$.

1. For the case $w_{0\varepsilon}'(\psi) \geqslant 0$ the estimate (3.6) is valid with the constant $M_2 = w_{0\varepsilon}(\frac{1}{\varepsilon})$, i.e. the following inequalities hold

$$\varepsilon \leqslant w(x,\psi,\varepsilon) \leqslant w_{0\varepsilon}(\tfrac{1}{\varepsilon}), \quad (x,\psi) \in D_\varepsilon$$

Since $\varepsilon = \min_{\bar{D}_\varepsilon} w_\varepsilon$, $w_{0\varepsilon}(\frac{1}{\varepsilon}) = \max_{\bar{D}_\varepsilon} w_\varepsilon$, and $w'_{0\varepsilon}(\psi) \geqslant 0$, $\psi \in [0,\frac{1}{\varepsilon}]$, then the function $Z = w_{\varepsilon\psi}$ is nonnegative at the boundary ∂D_ε of the domain D_ε. In the domain D_ε Z satisfies the parabolic equation with bounded coefficients and in view of generalized principle of maximum $Z \geqslant 0$, $(x,\psi) \in D_\varepsilon$.
2. Since $w_x(x,0,\varepsilon) = w_x(x,\frac{1}{\varepsilon},\varepsilon) = 0$ and

$$w_x\big|_{x=0} = \mu\sqrt{w_{0\varepsilon}(\psi)w''_{0\varepsilon}} - \varphi_0(x|w_0) < 0$$

with sufficiently small $\varepsilon \leqslant \varepsilon_0$, the function $\sigma(x,\psi) = w_{\varepsilon x}$ $(\varepsilon \leqslant \varepsilon_0)$ is nonnegative at the boundary ∂D_ε of the domain D_ε and satisfies in the domain D_ε the parabolic equation with bounded coefficients and nonpositive free term and, consequently, $\sigma \leqslant 0$ with $(x,\psi) \in D_\varepsilon$.

Passing to the limit as $\varepsilon \to 0$ in the resulting inequalities for w_ε we shall arrive to the claim of the theorem (3.2).

REFERENCES

1. Korteweg, D.J. (1901) Sur la forme que prennent les equations du mouvement des fluides si l'on tient compte des forces capillaires par des variations de densite, Archives Nederlandaises des Sciences Exactes and Naturelles, Series II, 6, p.1-24.

2. Slemrod, M. (1983) Admissibility criteria for propagating phase boundaries in a van der Waals fluid, Arch. Ration. Mech. Anal., v.81, No.1, pp.301-315.

3. Antontsev, S.N., Kazhikhov, A.V., Monakhov, V.N. (1990) Boundary value problems in mechanics oi homogeneous fluid. Studies in Mathematics and Their Applications, 22, North-Holland.

4. Polubarinova-Kochina, P.Ya. (1962) Theory of Ground Water Movement, English version Princeton Univ.Press, Princeton, N.J.

5. Oleinik, O.A. (1963) On a system of boundary layer theory. English version in USSR Comput. Math. and Math. Phys., 3.

6. Khusnutdinova, N.V. (1990) On the structure of the domain of degeneracy of a system of quasilinear parabolic equations. English version in Sov.Math.Dokl., v.40, N1.

7. Khusnutdinova, N.V. (1990) One model of a boundary layer with separation, Dinamika Sploshnoi Sredy, V.97, p.146-166, in Russian.

ASYMPTOTIC STABILITY FOR NONLINEAR PARABOLIC SYSTEMS

PATRIZIA PUCCI
University of Perugia, Italy

JAMES SERRIN
University of Minnesota

1. Introduction

The problem of asymptotic stability for second order ordinary differential equations is well–known in the literature. Recently, various extensions of this work have been given for the case of second order hyperbolic systems (see [1–5]). On the other hand, the situation for parabolic systems has received much less discussion, so that a study of this problem seems worthwhile.

We consider specifically systems of the form

$$\begin{cases} A(t)|u_t|^{m-2}u_t = \Delta u - f(x,u), & (t,x) \in J \times \Omega, \\ u(t,x) = 0, \quad (t,x) \in J \times \partial\Omega, \end{cases} \tag{1.1}$$

where $J = [0,\infty)$ and Ω is a bounded open subset of $\mathbb{R}^n$. The values of u are taken in $\mathbb{R}^N$, $N \geq 1$, and

$$A \in C(J \to \mathbb{R}^{N\times N}), \qquad f \in C(\Omega \times \mathbb{R}^N \to \mathbb{R}^N),$$

with $m > 1$ a fixed exponent. In order that (1.1) be parabolic, it is necessary to have

$$\big(A(t)v, v\big) > 0 \qquad \text{for all } t \in J \text{ and } v \in \mathbb{R}^N \setminus \{0\}, \tag{1.2}$$

where $(\cdot,\cdot)$ is the inner product in $\mathbb{R}^N$; in particular note that A need not be symmetric. Moreover, we assume that f represents a *restoring force* derivable from a potential F, namely,

$$\big(f(x,u), u\big) \geq 0 \tag{1.3}$$

and

$$f(x,u) = \frac{\partial F}{\partial u}(x,u), \tag{1.4}$$

where $F \in C^1(\Omega\times\mathbb{R}^N \to \mathbb{R})$ and $F(x,0) = 0$. The last condition is a normalization which can be assumed without loss of generality.

Note specifically that the relation (1.4) is redundant when $N = 1$, since in this case F can be obtained simply by integrating the force with respect to u.

S. N. Antontsev et al. (eds.), Energy Methods in Continuum Mechanics, 66–74.

If we take $m = 2$ then (1.1) reduces to the strongly coupled parabolic system

$$A(t)u_t = \Delta u - f(x, u),$$

or simply to

$$u_t = \Delta u - f(x, u),$$

when $A(t) = I$.

A canonical example of the type of functions f which we contemplate here is

$$f(x, u) = V(x)|u|^{p-2}u, \tag{1.5}$$

where $1 < p < \infty$, $m \leq \max\{p,\ 2n/(n-2)\}$ and $V \in C(\overline{\Omega} \to \mathbb{R}_0^+)$.

In the context of problem (1.1) the question of asymptotic stability of the rest state is best considered in terms of the natural energy associated with solutions of the system, namely,

$$Eu(t) = \int_\Omega \{\tfrac{1}{2}|Du(t,x)|^2 + F(x, u(t,x))\}dx.$$

In particular, the rest field $u(t,x) \equiv 0$ will be called *asymptotically stable (in the mean)*, if and only if

$$\lim_{t\to\infty} Eu(t) = 0 \qquad \text{for all solutions } u = u(t,x) \text{ of (1.1).}$$

In this formulation we have tacitly assumed that solutions are classical, but for a useful theory one must actually consider solutions in a wider class of functions. We treat this question in Section 2. Our main result is formulated in Section 3, and several applications are given in Section 4, including for example the mean curvature equation

$$A(t)u_t = \operatorname{div}\left(\frac{Du}{\sqrt{1 + |Du|^2}}\right) \tag{1.6}$$

and the degenerate Laplace system

$$A(t)u_t = \operatorname{div}(|Du|^{s-2}Du), \qquad s > 1. \tag{1.7}$$

We assume throughout that the reader is familiar with the results of reference [5], particularly Sections 2, 3 and 6.

The linear case

An important special case of (1.1) and (1.5) occurs when $p = m = 2$ and $A(t) = a(t)t^\alpha$, with $a \in C(J \to \mathbb{R}_0^+)$ and $\alpha \in \mathbb{R}$. The system then becomes (with $N = 1$)

$$a(t)t^\alpha u_t = \Delta u - V(x)u.$$

If a is bounded on J, and V is as above, then we prove that the rest state is asymptotically stable if $\alpha \le 1$. On the other hand, if a is also *bounded from zero* in J, then for $\alpha > 1$ there exist solutions which approach non-zero functions $\psi = \psi(x)$ as $t \to \infty$.

2. Definition of Solutions

To provide an appropriate definition for solutions of (1.1) it is convenient first to introduce the elementary bracket pairing in $\Omega \subset \mathbb{R}^n$

$$\langle \varphi, \psi \rangle = \langle \varphi, \psi \rangle(t) = \int_\Omega (\varphi, \psi) dx, \qquad \text{where } \varphi,\ \psi : J \times \Omega \to \mathbb{R}^N,$$

this being a well-defined real function of time for all $t \in J$ such that $(\varphi, \psi) \in L^1(\Omega)$. We write for simplicity

$$L^p = \left[L^p(\Omega)\right]^N, \qquad X = \left[W_0^{1,2}(\Omega)\right]^N,$$

where $p > 1$; these spaces are endowed respectively with the natural norms

$$\|\varphi\|_{L^p}, \qquad \|\varphi\|_X = \|D\varphi\|_{L^2} = \|D\varphi\|.$$

Now define $K' = C(J \to X)$ and

$$K = \{\phi \in K' \ : \ E\phi \text{ is locally bounded on } J\},$$

where $E\phi$ is the *total energy of the field* ϕ, that is

$$E\phi = E\phi(t) = \tfrac{1}{2}\|D\phi\|^2 + \int_\Omega F(x, \phi(t,x)) dx,$$

it being tacitly assumed that $F(\cdot, \phi(t, \cdot)) \in L^1_{\text{loc}}(\Omega)$ for all $t \in J$.

We can now give our principal definition. A *strong solution of* (1.1) is a function $u \in K$ which is weakly differentiable with respect to t in $J \times \Omega$ and which satisfies the following *two conditions*:

(A) *Conservation Law*

$$\mathcal{R}(t) \in L^1_{\text{loc}}(J), \tag{i}$$

$$Eu\Big]_0^t = -\int_0^t \mathcal{R}(s)ds \quad \text{for all } t \in J, \tag{ii}$$

where $\mathcal{R}(t) = \langle A(t)u_t, |u_t|^{m-2}u_t\rangle$.

(B) *Distribution Identity*

$$\int_0^t \{\langle Du, D\phi\rangle + \langle A(s)u_t, |u_t|^{m-2}\phi\rangle + \langle f(\cdot,u),\phi\rangle\}ds = 0$$

for all $t \in J$ and $\phi \in K$.

For various comments on this definition in the related context of damped wave systems, see [5].

An important issue is to determine a class of functions $A(t)$ and $f(x,u)$ for which the second and third terms in the distribution identity are well–defined, i.e. satisfy

$$\langle A(t)u_t, |u_t|^{m-2}\phi\rangle,\ \langle f(\cdot,u),\phi\rangle \in L^1_{\text{loc}}(J). \tag{2.1}$$

Letting $r = 2n/(n-2)$ be the Sobolev exponent for the space X ($r = \infty$ if $n = 1$; $2 < r < \infty$ if $n = 2$, since Ω is bounded), we make the natural hypothesis

$$|f(x,u)| \le \text{Const.}\,[1 + |u|^{p-1}], \qquad p > 1. \tag{2.2}$$

Moreover, if $n \ge 3$ and $p > r$, we suppose there are constants $\kappa > 0$ and $\kappa_0 \ge 0$ such that

$$\big(f(x,u),u\big) \ge \kappa|u|^p - \kappa_0|u|. \tag{2.3}$$

Also, for $t \in J$, let

$$H(t) = |A(t)|, \qquad h(t) = \min_{|v|=1}\big(A(t)v,v\big); \tag{2.4}$$

by (1.2) it is clear that both $h(t)$ and $H(t)$ are positive and continuous on J.

Under assumptions (2.2)–(2.3), the conditions (2.1) then hold provided that

$$m \le \max\{p,\ r\}, \qquad \delta \in L^1_{\text{loc}}(J), \tag{2.5}$$

where $\delta = \delta(t) = H^m/h^{m-1}$. This is proved almost exactly as in Section 2 of [5]. In particular, Lemmas 2.1 and 2.2 of [5] hold exactly as stated, so that $\langle f(\cdot,u),\phi\rangle$ is locally bounded on J when $u,\ \phi \in K$. Hence this term in (B) is well–defined.

To show that the first part of (2.1) also holds, we observe by Schwarz' inequality and (2.4) that, for $(t,x) \in J \times \Omega$,

$$\begin{aligned}|\big(A(t)u_t, |u_t|^{m-2}\phi\big)| &\le H(t)|u_t|^{m-1}|\phi| \le H(t)|u_t|^{2/m'}|u_t|^{(m-2)/m'}|\phi| \\ &\le H(t)\,h(t)^{-1/m'}\big(A(t)u_t,u_t\big)^{1/m'}|u_t|^{(m-2)/m'}|\phi| \\ &= \delta(t)^{1/m}\big(A(t)u_t,|u_t|^{m-2}u_t\big)^{1/m'}|\phi|,\end{aligned} \tag{2.6}$$

where m' is the Hőlder conjugate of m. In turn, from Hőlder's inequality and the definition of $\mathcal{R}(t)$,

$$|\langle A(t)u_t, |u_t|^{m-2}\phi\rangle| \le \delta(t)^{1/m} \cdot \mathcal{R}(t)^{1/m'} \cdot \|\phi\|_{L^m}, \tag{2.7}$$

for $t \in J$. Hence by Hőlder's inequality again, together with the facts that $\delta(t)$ and $\mathcal{R}(t)$ are in $L^1_{\mathrm{loc}}(J)$, and that $\|\phi\|_{L^m}$ is in $L^\infty_{\mathrm{loc}}(J)$, see Lemma 2.1 of [5], the condition $(2.1)_1$ follows at once.

Remark. When $N = 1$, or when $A(t)$ is a multiple of the identity matrix, we have $h(t) = H(t) = |A(t)|$, so that $\delta(t) = |A(t)|$. Moreover, if either $n = 1$ or 2, the restriction on m reduces simply to $1 < m < \infty$.

3. Asymptotic Stability

We can now give the main result of the paper.

THEOREM 3.1. *Let* (2.2), (2.3) *and* (2.4) *hold. Suppose there exists a non-negative continuous function* k *on* J, *such that*

$$k \notin L^1(J), \tag{3.1}$$

$$\liminf_{t\to\infty} \int_0^t \delta\, k^m ds \Big/ \left(\int_0^t k ds\right)^m < \infty. \tag{3.2}$$

Then the rest state $u \equiv 0$ *of* (1.1) *is asymptotically stable.*

Proof. The proof is essentially the same as for Theorem 3.1 of [5], with the main difference that (3.18) in that paper is here replaced by the relation

$$\int_T^t \big(\|Du\|^2 + \langle f(\cdot,u),u\rangle + \langle A(s)u_t, |u_t|^{m-2}u\rangle\big)k(s)ds = 0. \tag{3.3}$$

Because of the simpler form of this identity, it is no longer necessary to assume that k is of bounded variation, or to make use of the analogue of Lemma 3.3 of [5]. On the other hand, the analogues of Lemmas 3.1 and 3.4 continue to hold, as is easily seen. Moreover, for the analogue of Lemma 3.2 we have specifically, see (2.7),

$$\begin{aligned}\int_T^t |\langle A(s)u_t, |u_t|^{m-2}u\rangle| k(s)ds &\le \left(\int_T^t \delta\, k^m ds\right)^{1/m} \left(\int_T^t \mathcal{R}(s)ds\right)^{1/m'} \|u\|_{L^m}\\ &\le \varepsilon(T)\left(\int_0^t \delta\, k^m ds\right)^{1/m}\end{aligned}$$

for $t \geq T \geq 0$, and $\varepsilon(T) \to 0$ as $T \to \infty$, as in [5]. This being the case, we get from (3.3), exactly as in the proof of Theorem 3.1 of [5],

$$-\alpha \int_T^t k ds + \varepsilon(T) \left(\int_0^t \delta\, k^m ds \right)^{1/m} \geq 0$$

for $t \geq T \geq 0$. By (3.2) there is a sequence $t_i \nearrow \infty$ and a number $M > 0$ such that

$$\int_0^{t_i} \delta\, k^m ds \leq \left(M \int_0^{t_i} k ds \right)^m .$$

Consequently, taking T so large that $\varepsilon(T)M \leq \alpha/2$, we obtain

$$\frac{\alpha}{2} \int_0^T k ds - \frac{\alpha}{2} \int_T^{t_i} k ds \geq 0,$$

which yields an immediate contradiction with (3.1) when $t_i \nearrow \infty$. This completes the proof.

4. Applications

Important special cases of Theorem 3.1 occur when $k = 1$ and $k = \delta^{1/(1-m)}$. In the first case, the assumptions (3.1) and (3.2) reduce to the single condition

$$\liminf_{t \to \infty} \frac{1}{t^m} \int_0^t \delta(s) ds < \infty, \tag{4.1}$$

and in the second to

$$\int_0^\infty \delta^{1/(1-m)} dt = \infty. \tag{4.2}$$

An interesting extension of the results occurs if the term $A(t)|u_t|^{m-2}u_t$ in (1.1) is replaced by a general function of the form $Q(t, x, u, u_t)$. In this case it is necessary to define $\mathcal{R}(t) = \langle Q(t, \cdot, u, u_t), u_t \rangle$ in condition (A) of Section 2 and to replace the principal conditions (2.4), (2.5) by – see (2.6) –

$$|Q(t, x, u, v)| \leq \hat{\delta}(t)^{1/m} \cdot \big(Q(t, x, u, v), v\big)^{1/m'}, \tag{4.3}$$

where $1 < m \leq \max\{p,\ r\}$ and $\hat{\delta} \in L^1_{\text{loc}}(J)$. Then Theorem 3.1 continues to hold, provided in (3.2) we change $\delta\, k^m$ to $\hat{\delta}\, k^m$.

Finally, the Laplace operator in (1.1) can be replaced by various elliptic operators without affecting the results, as shown in Section 6 of [5]. Cases of particular importance arise for the mean curvature equation (1.6) and the degenerate Laplace system (1.7).

In the case of (1.6), for instance, we take $X = W_0^{1,1}(\Omega)$. Condition (2.2) holds trivially since $f \equiv 0$, while on the other hand (2.3) fails for all $p > 1$. It is therefore necessary to replace the first condition of (2.5) by $m \leq r$, where r is the Sobolev exponent for $W_0^{1,1}(\Omega)$, namely $r = n/(n-1)$. Obviously $m = 2$ in the present case, which gives the condition $2 \leq n/(n-1)$. This means that our stability results apply to (1.6) exactly when $n = 1$ or $n = 2$. More precisely, we have the following conclusion.

THEOREM 4.1. *Let u be a strong solution of equation* (1.6), *and suppose either* $n = 1$ *or* $n = 2$. *Assume there exists a non-negative continuous function k on J such that* (3.1) *and* (3.2) *hold with* $m = 2$ *and* $\delta(t) = A(t)$. *Then*

$$\lim_{t \to \infty} \|Du\|_{L^1} = 0. \tag{4.4}$$

Proof: Equation (1.6) corresponds to the case

$$A(w) = w/\sqrt{1+|w|^2}, \qquad G(w) = \sqrt{1+|w|^2} - 1, \qquad s = 1$$

in [5], Section 6.3. Since $f \equiv 0$, the total energy of the field u is easily seen to be given by

$$Eu(t) = \int_\Omega G(Du(t,x))\,dx.$$

As in Section 3 we then obtain $Eu(t) \to 0$ as $t \to \infty$, which is equivalent to the stated conclusion. This result also implies that $\|u\|_{L^2} \to 0$ as $t \to \infty$.

A similar argument also applies to the system (1.7), giving the corresponding

THEOREM 4.2. *Let u be a strong solution of the system* (1.7), *where*

$$s \geq 2n/(n+2).$$

Assume there exists a non-negative continuous function k on J such that (3.1) *and* (3.2) *hold with* $m = 2$ *and* $\delta(t) = H^2/h$. *Then*

$$\lim_{t \to \infty} \|Du\|_{L^s} = 0. \tag{4.5}$$

When $A(t)$ is uniformly bounded on J, the conditions (3.1) and (3.2) are satisfied with $k = 1$. In particular, then, the rest state for (1.6) and for (1.7) is asymptotically stable in the classical case $A(t) \equiv I$, provided that $n = 1$ or $n = 2$ for (1.6), or $s \geq 2n/(n+2)$ for (1.7).

Another case of interest occurs for the modified equations

$$\frac{u_t}{\sqrt{1+|Du|^2}} = \operatorname{div}\left(\frac{Du}{\sqrt{1+|Du|^2}}\right) \tag{1.6$'$}$$

and

$$\frac{u_t}{\sqrt{1+|Du|^2}} = \operatorname{div}(|Du|^{s-2}Du) \tag{1.7$'$}$$

(with $N = 1$), whose left hand sides represent the normal velocity of the surface $u = u(t,x)$ as a function of time in the $(n+1)$-dimensional (x,u) space. These equations cannot be put in the form (1.6), (1.7) because the coefficient $1/\sqrt{1+|Du|^2}$ depends on both t and x. On the other hand, by taking

$$Q(t,x,u,v) = \frac{v}{\sqrt{1+|Du(t,x)|^2}},$$

as at the beginning of this section, we find that

$$|Q(t,x,u,v)| \le \big(Q(t,x,u,v)v\big)^{1/2}.$$

Thus (4.3) holds with $m = 2$ and $\hat{\delta} \equiv 1$. In turn, the conclusion (4.4) of Theorem 4.1 holds for $(1.6)'$ when $n = 1$ or $n = 2$, and the conclusion (4.5) of Theorem 4.2 holds for $(1.7)'$ when $s \ge 2n/(n+2)$.

The linear case

Consider the problem

$$\begin{cases} a(t)t^\alpha u_t = \Delta u - V(x)u & \text{in } J \times \Omega, \\ u(t,x) = 0 & \text{on } J \times \partial\Omega, \end{cases} \tag{4.6}$$

where $N = 1$ for simplicity, Ω is a bounded open subset of $\mathbb{R}^n$, and $a \in C(J \to \mathbb{R}_0^+)$, $V \in C(\overline{\Omega} \to \mathbb{R}^+)$. Since $N = 1$ we have $\delta(t) = H(t) = a(t)t^\alpha$ in (2.5); it is convenient here to take $J = [1,\infty)$ in order to avoid the singularity at $t = 0$ when $\alpha < 0$. Then, with $m = p = 2$, it is clear that (2.2) and $(2.5)_1$ are satisfied. Moreover, assuming that

$$a(t) \le C \qquad \text{in } J, \tag{4.7}$$

we get $(2.5)_2$.

It follows now, either from (4.1) or (4.2), that *the rest state is asymptotically stable for* (4.6) *whenever* (4.7) *holds and* $\alpha \le 1$. [Actually, using (4.2), we find that asymptotic stability holds for $\alpha \le 1$ even when (4.7) is replaced by $a(t) \le C \log t$.]

When $\alpha > 1$, neither (4.1) nor (4.2) applies. In fact, in this situation solutions of (4.6) *do not* in general approach zero as $t \to \infty$. To illustrate this case, let φ_k be the kth eigenfunction of $-\Delta + V(x)$ in Ω, with Dirichlet boundary conditions. We say that a function

$$\psi = \psi(x) \in Y \equiv \operatorname{span}\{\varphi_k\}_{k=1}^\infty$$

is *attainable* if there exists a solution $u \in K$ of (4.6) such that

$$\lim_{t\to\infty} \|u(t) - \psi\|_{L^2} = 0. \tag{4.8}$$

THEOREM 4.3. *Suppose $\alpha > 1$ and also that $a(t) \geq 1/C$ for all $t \in J$. Then every function $\psi \in Y$ is attainable for problem (4.6). In turn, the set of attainable functions is dense in L^2.*

Proof: We first show that every eigenfunction φ_k is attainable. For this purpose consider the function

$$u_k(t, x) = w_k(t)\varphi_k(x),$$

which satisfies (4.6) if and only if w_k is a solution of the ordinary differential equation

$$a(t)t^\alpha w' + \mu_k w = 0, \qquad t \in J, \tag{4.9}$$

where $\mu_k > 0$ is the eigenvalue associated to φ_k. By integration we get

$$w(t) = \text{Const.} \exp\left(-\int_1^t \frac{\mu_k}{a(s)s^\alpha} ds\right).$$

Since $\alpha > 1$ and $a(s) \geq 1/C$ in J, the integral is convergent, whence

$$\lim_{t\to\infty} w(t) \quad \text{exists and is finite.}$$

It follows that the set of attainable limits of solutions of (4.9) is all of $\mathbb{R}$. Hence for the particular solution w_k of (4.9) which has limit value one at infinity, we get

$$\lim_{t\to\infty} \|u_k(t, \cdot) - \varphi_k(\cdot)\|_{L^2} = 0.$$

Finally, using the linearity of (4.6), we obtain (4.8) for every $\psi \in Y$.

Acknowledgment. P. Pucci is a member of *Gruppo Nazionale di Analisi Funzionale e sue Applicazioni* of the *Consiglio Nazionale delle Ricerche*. This research has been partly supported by the Italian *Ministero della Università e della Ricerca Scientifica e Tecnologica*.

Bibliography

1. A. Haraux, *Recent results on semilinear hyperbolic problems in bounded domains*, in *Partial Differential Equations*, Lecture Notes in Math., vol. **1324**, 118–126, Springer–Verlag, Berlin – New York, 1988.
2. P. Marcati, *Decay and stability for nonlinear hyperbolic equations*, J Diff. Equations **55** (1984), 30–58.
3. P. Marcati, *Stability for second order abstract evolution equations*, Nonlinear Anal. **8** (1984), 237–252.
4. M. Nakao, *Asymptotic stability for some nonlinear evolution equations of second order with unbounded dissipative terms*, J. Diff. Equations **30** (1978), 54–63.
5. P. Pucci & J. Serrin, *Asymptotic stability for non-autonomous dissipative wave systems*, submitted for publication, 37 pages.

NONLOCAL SYMMETRIES IN NONLINEAR HEAT EQUATIONS

V.V. PUKHNACHOV
Lavrentyev Institute of Hydrodynamics
Lavrentyev Prospect 15, Novosibirsk, 630090, Russia

Abstract. The paper is devoted to the study of symmetry properties of second-order nonlinear parabolic equations in divergent form and with a single space variable. It contains also a review of preceding results in the field, including the classical symmetries of the equations mentioned. The main topic of the paper are nonlinear symmetries of these equations.

1. Introduction

As is observed in [1, 3], any evolution equation of divergent form can be interpreted as the mass conservation law of the motion of a fictious liquid. This interpretation is facilitated by introduction of Lagrangian coordinates. Being nonlocal, this transformation allows one to find out hidden symmetries of the equations under consideration [4, 5]. Combinations of the classical and nonlocal symmetries make it possible to construct new exact solutions to nonlinear parabolic equations. A number of such solutions are presented in section 2 where the equation

$$u_t = [\varphi(u, u_x)]_x \,, \tag{1.1}$$

and its partial case, the equation

$$u_t = [k(u)u_x]_x \,, \tag{1.2}$$

are considered from the point of view of their group properties.

As is shown in [6, 7], the passage in (1.1) to the new independent variable ξ replacing x and the new sought function ω, replacing $u(x,t)$, by the formulas

S. N. Antontsev et al. (eds.), Energy Methods in Continuum Mechanics, 75–99.

$$\left\{ \begin{array}{l} \xi = \int_0^x u(y,t)dy + \int_0^t \varphi\left[u(0,s), u_x(0,s)\right] ds, \\ \omega(\xi,t) = \left[u(x,t)\right]^{-1} \end{array} \right. \tag{1.3}$$

leads to the following equation for function ω

$$\omega_t = \left[-\omega\varphi\left(\omega^{-1}, -\omega^{-3}\omega_\xi\right)\right]_\xi . \tag{1.4}$$

Following [4, 5], we can interpret the new space variable ξ as the mass Lagrangian coordinate and term the change of variables (1.3) *L-transformation.* This transformation establishes the correspondence between equation (1.1) and equation (1.4) of the same class by the rule

$$\varphi(u,p) \mapsto \psi(u,p) \equiv -u\varphi\left(u^{-1}, -u^{-3}p\right).$$

The last relation implies the equality

$$-u\psi(u^{-1}, -u^{-3}p) = \varphi(u,p),$$

which means that L-transformation is *involute* i.e. $L^2 = I$. The explicit form of the transformation inverse to (1.8) is

$$\left\{ \begin{array}{l} u(x,t) = \left[\omega(\xi,t)\right]^{-1}, \\ x = \int_0^\xi \omega(\eta,t)d\eta - \int_0^t \omega(0,s)\varphi\left[1/\omega(0,s), -\omega_\xi(0,s)/\omega^3(0,s)\right] ds. \end{array} \right. \tag{1.5}$$

It may happen, however, that equation (1.4), having been subjected to L-transformation, coincides with the origin. These equations we will term *L-invariant.* Comparing (1.1) and (1.4) we get the conditions of L–invariance

$$\varphi(u,p) = -u\varphi(u^{-1}, -u^{-3}p), \tag{1.6}$$

which can be viewed as a functional equation for the function φ. Several classes of partial solutions to (1.6) containing arbitrary functions of a single variable are discovered in [5, 8]. The general solution of equation (1.6) for positive u is presented in [9]. Section 3 is devoted to Lagrange-invariant equations.

It is worth noting that the considerations of papers [1—9] referred to the equations in which the space variable did not enter explicity. In the conclusion of section 4 we present examples of nonlocal invariant parabolic equations with spherical and cylindrical symmetries.

2. Hidden Symmetry of Parabolic Equations.

Let us consider first the nonlinear heat-conduction equation (1.2). (When saying "nonlinear heat conductivity" we mean that the coefficient of heat conductivity $k(u) \neq const$). It is makes sense to pose for this equation, including an "arbitrary" element $k(u)$, the problem of the group classification: to separate from all equations of the type (1.2) those whose admissible Lie group is wider than for the equation of the general form. Note that the transformation of the type $E = E(a, b, c)$

$$\tilde{u} = au + b, \quad \tilde{k} = ck, \quad a, b, c = const., \quad ac \neq 0, \tag{2.1}$$

renders equation (1.2) an equivalent one but with another function k. Equations connected by poitwise equivalence transformations possess Lie groups similar to $E(a, b, c)$ and in the group classifications are considered undistinguible.

The problem of the group classification of equation (1.2) was posed and solved in 1959 by L.V.Ovsyannikov [10] (see also monograph [11] where the below-utilized termilogy of the group analysis of differential equations is introduced). Let us formulate the main result of paper [10]. If $k \neq const$, $k \neq e^u$, $k \neq u^m$, then the main group admitted by equations (1.2) is three-parameter. The basis of the corresponding Lie algebra is generated by the infinitesimal operator

$$X_1 = \partial/\partial x, \quad X_2 = \partial/\partial t, \quad X_3 = 2t\partial/\partial t + x\partial/\partial x. \tag{2.2}$$

If $k = e^u$, then the group extends up to four-parameter on account of the operator

$$X_4 = t\partial/\partial t - \partial/\partial u. \tag{2.3}$$

If $k = u^m$, $m \neq -4/3$, then the main group of equation (1.2) is also four-parameter. The basis of the Lie algebra here is constituted by the operators (2.2) and

$$X_5 = mx\partial/\partial x + 2u\partial/\partial u. \tag{2.4}$$

Lastly, in the exeptional case $k = u^{-4/3}$ equation (1.2) admits the five-parameter group of pointwise transformations generated by operators (2.2), (10.6) with $m = -4/3$, and the additional operator

$$X_6 = -x^2\partial/\partial t + 3xu\partial/\partial u. \tag{2.5}$$

The foregoing constructions are framed by the following observation: besides pointwise transformations (2.1), equation (1.2) and more general equation (1.1) admit a nonlocal equivalence transformation which is realized via the passage to Lagrangian coordinates (L-transformation). The explicit form of L-transformation of equation (1.1) is given by formulas (1.3) and its specification for the case $\varphi(u,p) = k(u)p$ is the following:

$$\begin{cases} \xi = \int_0^x u(y,t)dy + \int_0^t k\left[u(0,s)\right] u_x(0,s)ds, \\ \tau = t, \quad \omega(\xi,\tau) = \left[u(x,t)\right]^{-1}; \end{cases} \tag{2.6}$$

it is assumed here that $u \neq 0$. Since transformation (1.3) is not pointwise, the Lie groups admitted by the original and transformed equations need not be isomorphic. This feature enforces one to construct equations (1.1) for which the counterparting equation (1.4) possesses a wider Lie group. An example of this situation is delivered by the nonlinear heat-conduction equation under an appropriate choice of the function $k(u)$. So, the equation

$$u_t = \left[u^{-2} \exp\left(u^{-1} u_x\right)_x\right] \tag{2.7}$$

admits a three-parameter Lie group [10]. Had we subject it to L-transformation, we would get the equation

$$\omega_\tau = \left(e^\omega \omega_\xi\right)_\xi, \tag{2.8}$$

which possesses the four-parameter main group.

The group of pointwise transformations of the equation

$$u_t = \left(u^{-2/3} u_x\right)_x \tag{2.9}$$

is three-parameter. In the result of L-transformation we obtain the equation

$$\omega_t = \left(\omega^{-4/3} \omega_\xi\right)_\xi, \tag{2.10}$$

which has the widest (five-parameter) main group among all nonlinear equations of the type (1.2). Having rewritten in the new notation the additional operator of the Lie algebra of this group, we arrive at the corresponding finite transformation

$$\tilde{\tau} = \tau, \quad \tilde{\xi} = (1 + \alpha\xi)^{-1}, \quad \tilde{\omega} = (1 + \alpha\xi)^3 \omega \tag{2.11}$$

with a parameter $\alpha \in \mathbf{R}$, which conserves equation (2.10). There are no finite transformations in the space (x,t,u), admitted by equation (2.9), which correspond to transformation (2.11).

Let us denote by U the manyfold in $\mathbf{R}^3$ determined by a solution $u = u(x,t)$ of equation (2.9), and by Ω the corresponding manyfold $\omega = \omega(\xi,\tau)$.

The mapping of U into Ω realized by the change of variables (2.6) we denote by $\Omega = LU$, and mapping (2.11) by $\overline{\Omega} = T_\alpha \Omega$. Let us define the family of transformations S_α by the formula

$$S_\alpha U = L^1 \overline{\Omega} \equiv L^{-1} T_\alpha L U.$$

(Due to (2.6), the invertibility of L-transformation is garanteed by the condition $u \neq 0$). It follows from (2.11) and the definition of S_α that

$$\forall \alpha, \beta \in \mathbf{R} \quad S_\alpha S_\beta U = S_{\alpha+\beta} U; \quad \text{besides, } S_0 U = U.$$

Thus, the set of transformations S_α form a one-parameter group which, obviously, is not the Lie group.

The transformation S_α maps each of solutions of equation (2.9) again into a solution of this equation. This property may be termed *the hidden symmetry* of equation (2.9). An analogous property of hidden symmetry is also intrinsic for equation (2.7). Along with (2.11) one should make use of the transformation

$$\tilde{\xi} = \xi, \quad \tilde{\tau} = \tau \exp(\alpha), \quad \tilde{\omega} = \omega - \alpha,$$

generated by operator (2.3) and conserving equation (2.7), reciprocical to (2.8).

Remark that both equations (2.7) and (2.8) are not outlined by the groups they admit and even do not have a nontrivial Lie-Bäcklund group, [12]. The above-discovered invariance of these equations with respect to nonlocal transformations S_α is of another nature.

The term "hidden symmetry" (as well as the encompassing ones "nonlocal" and "non-pointwise" symmetries) were recently given certain fame due to different methods of finding nonlocal symmetries of evolution equations which cannot be found in the borders of the classical Lie approach (see, for instance [4, 5, 13, 14, 15, 16, 17] and references therein).

Apparently, the first example of this sort was discovered by V.A.Fok [18] in 1935. It was established that after the Fourier transformation the stationary Schrödinger equation for the hydrogen atom admitted the group of rotations in the four-dimensional space. Later on, the hidden symmetry of a number of linear equations of mathematical physics, (particularly, the Maxwell equation), was studied in the works by V.I.Fushchich, A.G.Nikitin and their collaborators (see [19, 20, 21] and the bibliography therein). The methods of finding hidden symmetries used in [18, 19, 20, 21] are specific for linear equations.

As far as nonlinear equations of the type (1.1) are concerned, besides (1.2), a fairly easy analysis may be given to the hidden symmetry properties of equations of the class

$$v_t = [K(v_x)]_x\,, \tag{2.12}$$

describing a shear flow of a plastic-viscous medium and processes of non-newtonian filtration. It is to be noted that in the cases when the function $K(u)$ is the primitive of the function $k(u)$ from equation (1.2), there exists a simple connection between solutions of equations (1.2) and (2.12): $u = v_x$. In this case the equation

$$v_t = k(v_x)v_{xx} \tag{2.13}$$

is an equivalent form of (2.12).

On the other hand, the substitution $v = w_x$ connects equation (2.12) to the equation

$$w_t = K\left(w_{xx}\right), \tag{2.14}$$

which, generally speaking, is neither divergent nor quasilinear. Thus, we have the sequence of three equations, (1.2), (2.12), and (2.14), connected one to another by the transformations $u = v_x = w_{xx}$. In the series of works [13, 17, 22] the group properties of equations of this sequence were studied and nonlocal symmettries were discovered admitted by its members under special choices of $K(u)$.

Let us return to equation (2.13), equivalent to (2.12). In the "general" case, this equation admits the four-parameter Lie group. If $k = \exp(v_x)$ and $k = v_x^m$, $m \neq -2$, then the main group of equation (2.13) is five-parameter. For $m = 0$ this equation is linear, in case $m = -2$ it can be linearized with the help of the so-called von Mises transformation. The problem of the group classification of the equation under discussion is solved in [23]. Besides the cases already pointed out, there is the exeptional dependence

$$k = \left(1 + v_x^2\right)^{-1} \exp\left(\nu \arctan v_x\right), \quad \nu \geq 0, \tag{2.15}$$

when the main group extends up to a five-parameter one. The additional operator of the Lie algebra of this group has the form

$$Y_5 = -\nu t\partial_t - v\partial_x + x\partial_v. \tag{2.16}$$

(It is curious to note that equation (2.13) in the complex domain, and with k given by formula (2.15), can be transformed into the equation with power nonlinearity $k = v_x^{m-1}$ [24]). As is shown in [13], in the exclusive case of (2.15) the transformations of the group generated by operator (2.16) and conserving equation (2.13) are not corresponded by any pointwise transformation admitted by two other equations (1.2) and (2.14) of the three-term sequence originated by the given function k. Since the transition from the function v to w is nonlocal, $v = w_x$, it follows, in particular, that the equations

$$w_t = \arctan(w_{xx}),$$

$$w_t = \nu^{-1}\exp(\nu\arctan(w_{xx})), \quad \nu > 0,$$

also possess hidden symmetry. These equations are implied by (2.14) after the specialization of k of the form (2.15) with the equality

$$\frac{dK(u)}{du} = k(u).$$

The case when the Lie group admitted by equation (1.2) can be extended in comparison with that of the corresponding equation (2.13) is exhausted by the dependence

$$k(u) = u^{-4/3}.$$

One another possibility when a nontrivial operator (i.e. that cannot be reduced to the appearance in the Lie algebra the operator $\partial/\partial w$) extending the main Lie group occurs in the passage from (2.13) to equation (2.14) be realized in the cases $k = u^{-4/3}$ and $k = u^{-2/3}$ only. Here equation (2.14) is equivalent to the equations

$$w_t = w_{xx}^{-1/3}, \tag{2.17}$$

$$w_t = w_{xx}^{1/3}, \tag{2.18}$$

respectively. Each of equations (2.17), (2.18) admits a seven-parameter group of pointwise transformations, [13]. Equations (2.17), (2.18) are also remarkable because the dimension of the Lie algebras they admit is the maximal among all equations not equivalent to the linear equation $u_t = F(t, x, u, u_x, u_{xx})$. This assertion is a very partial implication of results of B.A.Magadeev, [25], who solved the problem of the group classification of the equations

$$u_t = F\left(t, x, u, \frac{\partial}{\partial x}, \dots, \frac{\partial^n}{\partial x^n}\right)$$

with respect to contact transformations. Moreover, it follows from [16, 26, 27] that equation (2.18) can be rendered (2.17) by means of contact transformations. Not having a possibility to go deep into the theory of contact transformations we readdress the reader to monographs [16, 26, 27] and the survey paper [28] presenting the modern state of the subject.

So, we have at our disposal a few equations displaying the hidden symmetry property. Let us demonstrate how one may utilize this property to construct exact solutions of equations not invariant in the classical sense. For example, let us consider equation (2.10). It admits the infinitesimal operator

$$\Sigma = \partial_\tau - \xi^2 \partial_\xi + 3\xi\omega\partial_\omega .$$

One may take for the basis of invariants of this operator

$$I_1 = \xi^{-1} - \tau, \quad I_2 = \xi^3 \omega .$$

According to the techniques of [11], one has to find a solution of equation (2.10) invariant with respect to the operator Σ in the form

$$\omega = \xi^{-3} f\left(\xi^{-1} - \tau\right) . \tag{2.19}$$

The ordinary differential equation for the function f, following after the substitution of (2.19) into (2.10), can be integrated and then reduced to the form

$$\frac{df}{d\xi} = -f^{4/3}(f + C) , \tag{2.20}$$

where C is the constant of integration. For the sake of definiteness we suppose $C > 0$. All nonnegative solutions of equation (2.20) can be obtained by a transfer along the axis ξ from the single solution defined for $\xi > 0$ and satisfying the condition $f \to \infty$ as $\xi \to 0$. Not loosing generality we may thus suppose that the function $f(\xi)$ is defined by the relation

$$\xi = \int_f^\infty \frac{dg}{g^{4/3}(g + C)} .$$

The constructed solution $\omega = \omega_*(\xi, \tau)$ of equation (2.10) is defined in the domain

$$\Delta = \{\xi, \tau : \xi > 0, \tau > 0, \xi\tau > 0\} .$$

The function ω_* is positive in this domain and grows unlimitedly near that part of the boundary where $\xi\tau = 1$. The last feature does not allow one to use the explicit formulas (2.6) and the solution $u = u_*(x, t)$ of equation (2.9), reciprocical to (2.10). The requested representation of the solution

u_* is derived by the replacement of the loop of integration in the second line of formulas (1.5) by the curve connecting the points (ξ, t), $(\xi, 0)$, and $(\infty, 0)$, so that

$$x = -\int_{\xi}^{\infty} \omega(\eta, 0) d\eta + \int_{0}^{t} \omega^{-4/3}(\xi, s)\omega_{\xi}(\xi, s) ds.$$

Substituting here and into the first line of formulas (1.5) $\omega = \omega_*(\xi, t)$, after simple transformations we find the parameter representation for the solution u_* of equation (2.9):

$$\begin{cases} u_* = \xi^3 \left[f\left(\xi^{-1} - t\right)\right]^{-1}, \\ x = -\int_0^{\xi^{-1}} \xi f(\xi) d\xi + \int_{\xi^{-1}-t}^{\xi^{-1}} \left\{\xi^{-1}\left[f(\xi) + C\right] - 3f^{-1/3}(\xi)\right\} d\xi. \end{cases} \tag{2.21}$$

The integrals in (2.21) converge as $\xi \to 0$ and $\xi \to \infty$ for each $t \in [0, \xi^{-1})$. This follows from the asymptotic dependences

$$f = \left(\frac{3}{4\xi}\right)^{3/4}\left[1 + O\left(\xi^{3/4}\right)\right] \quad \text{as } \xi \to 0,$$

$$f = \left(\frac{3}{C\xi}\right)^{3}\left[1 + O\left(\xi^{-3}\right)\right] \quad \text{as } \xi \to \infty.$$

In particular, as $\xi \to 0$ we have

$$\lim_{\xi \to 0} \int_{\xi^{-1}-t}^{\xi^{-1}} \left[C\xi^{-1} - 3f^{-1/3}(\xi)\right] d\xi = \frac{Ct^2}{2}.$$

Thereby, the image of the ray $\{\xi = 0, \tau > 0\}$ under the mapping $(\xi, t) \mapsto (x, t)$ via L-transformation generated by equation (2.10) will be the parabola l_1 governed by the equation $x = x_1(t) \equiv -A + Ct^2/2$, $t > 0$ with

$$A = \int_0^{\infty} \xi f(\xi) d\xi.$$

Proceeding to the limit in the first line of (2.20), we get

$$u_* = \left(\frac{C}{3}\right)^3 \quad \text{as } x = x_1(t),\ t > 0. \tag{2.22}$$

Letting in (2.20) $t = 0$, we see that this transformation renders the ray $\{\xi > 0, \tau = 0\}$ the segment $l_0 = \{-A < x < 0, t = 0\}$. If function $\psi(x)$ is defined by the relation

$$x = -\int_0^{\psi} \xi f(\xi) d\xi \quad \text{for } x \in (-A, 0],$$

then, due to (2.21), the values of function u_* at the points of this segment can be calculated by the formula

$$u_*(x,0) \equiv u_0(x) = [\psi(x)]^{-3} \{f[\psi(x)]\}^{-1} \quad \text{as } -A < x < 0. \tag{2.23}$$

It follows that $u_0 \to (C/3)^3$ as $x \to -A$, i.e., at the point $x = -A$, $t = 0$ the compatibility condition of initial and boundary data (2.22), (2.23) is fulfilled. Besides, function u_0 grows unlimitedly with $x \to -0$. Moreover,

$$u_0 = (48)^{3/5} (-5x)^{-9/5} \left[1 + O\left(|x|^{3/5}\right)\right] \quad \text{as } |x| \to -0.$$

Now set in (2.20) $\xi = t^{-1}$. The equation

$$x = x_2(t) \equiv Ct^2 + \int_0^t \left[(t-\xi)f(\xi) - 3f^{-1/3}(\xi)\right] d\xi$$

defines on the plane (x,t), $t > 0$ a curve exiting from the point $(0,t)$. Using equation (2.20) and the asymptotics of function f as $\xi \to \infty$, we find: $x_2'(0) = 0$, $x_2'' = C$ for all $t > 0$. Thereby, $x_2 = Ct^2/2$, i.e., the curve l_2 is also a parabola. Besides, $u_*(x,t) \to 0$ as $x \to x_2(t) - 0$, $t > 0$. Thus, l_2 is a line of degeneracy of equation (2.9). It is corresponded by that part of the boundary $\xi\tau = 1$ of domain Δ where the function ω_* is defined. Hence, under the mapping $(\xi, \tau) \mapsto (x,t)$ the image D of the domain Δ is the domain bounded by parabolas l_1, l_2, and segment l_0.

Equation (2.9) belongs to the class of the so-called *fast diffusion equations*: if u is interpreted as the concentration of a substance and the equation itself is considered as a nonlinear diffusion equation, then the diffusion coefficient $u^{-2/3} \to \infty$ as $u \to 0$. (More information on the fast diffusion equation can be found in [29, Ch.2]; see also the references to this chapter).

Returning to the function u_*, we can view it as a solution of equation (2.9) in the domain D satisfying conditions (2.22), (2.23), and the condition

$$u_* \to 0 \quad \text{as } x \to x_2(t) - 0,\ t > 0. \tag{2.24}$$

It is easy to show that along with (2.24) the relations hold

$$\partial_t u_* \to 0, \quad \partial_x u_* \to 0, \quad \partial_{xx}^2 u_* \to 0 \quad \text{as } x \to x_2(t) - 0,\ t > 0.$$

A natural desire is to continue the function u_* by zero into the domain $\{x > x_2(t), t > 0\}$ so as to make the continued function (let us denote it U_*) a weak solution of equation (2.9) in the domain $D_+ = \{(x,t) : x > x_1(t), t > 0\}$. However, this is impossible because of the following reason: the diffusion flux $-u_*^{-2/3}\partial u_*/\partial x$ does not tend to zero as $x \to x_2(t) - 0$. Making use of the relations

$$\lim_{x \to x_2 - 0} u_*^{-2/3}\frac{\partial u_*}{\partial x} = - \lim_{\xi \to t^{-1}+0} \omega_*^{-7/3}\frac{\partial \omega_*}{\partial \xi}$$

and referring to (2.19)-(2.21), we find:

$$u_*^{-2/3}(u_*)_x \to -t^{-2} \quad \text{as } x \to x_2(t) - 0,\ t > 0. \tag{2.25}$$

Since the solution u_* of equation (2.9) satisfies relation (2.25) at the line of degeneracy, the function U_* obeys in the domain D_+ the equation

$$u_t = \left[u^{-2/3}u_x + t^{-2}H(x - x_2(t))\right]_x, \tag{2.26}$$

where $H(x)$ is the Heavyside function.

Apparently, equations similar to (2.26) have not been given yet a detailed mathematical treatment. In 1990, in a private communication J.L.Vazquez pointed out that one might interpret as solutions of equations of the type (2.26) the solutions of the fast diffusion equations with discontinuities of the diffusion flux at the line (or the surface) of degeneracy.

Besides the features outlined, a solution of the initial-boundary value problem (2.9), (2.22)-(2.24) possesses a number of additional interesting properties. The function $u_0(x)$ in the initial condition (2.23) has a strong singularity at the point $x = 0$ - this function even does not belong to the class $L_1(-A, 0)$. On the other hand, for each $t > 0$ the function u_* turns out to be smooth up to the line of degeneracy. Particularly, this implies that "the mass"

$$M(t) = \int_{x_1(t)}^{x_2(t)} u_*(x,t)dx,$$

associated with the solution of problem (2.9), (2.22)-(2.24), is finite for each $t > 0$ while $M \to \infty$ as $t \to 0$.

As was already mentioned, having continued u_* by zero into the domain $D_+ \setminus D$, we go out the class of generalized solutions to equation (2.9). There is another way of continuing a solution to this equation across the line of degeneracy, however the continued function fails to remain nonnegative. Let us make in equation (2.9) the substitution $u = z^3$. The function z is a solution of the equation

$$z^2 z_t = z_{xx} \quad \text{in } D. \tag{2.27}$$

According to (2.24), (2.25), this function satisfies at the line l_2 the conditions

$$z = 0, \quad z_x = -\frac{t^{-2}}{3} \quad \text{as } x = \frac{Ct^2}{2}, t > 0. \tag{2.28}$$

By the Cauchy-Kovalevskaya theorem, there exists a solution z_* of the Cauchy problem (2.27), (2.28) analitic in a neighborhood of the line l_2, $t > 0$. The function z_*^3 coincides with u_* in D_+ and is an analitic continuation of the solution u_* of equation (2.9) across the line of degeneracy for $x > x_2$, $t > 0$ ($x - x_2$ is small enough). By (2.28), it is clear that z_* (and along with it the continuation of the function u_*) assume negative values for small positive values of $x - x_2(t)$.

Solution (2.21) is an infrequent example of the exact solution to the fast diffusion equation, which is not a self-similar one. Moreover, this solution does not belong at all to the set of solutions of equation (2.9) invariant with respect to the group of pointwise transformations. (Let us remind that the generators of this group are the operators X_1, X_2, X_3 (2.2), and X_5 (2.4)). To prove this assertion one has to subject the function $\Phi = u - u_*(x, t)$ to the operator

$$X = \sum_{j=1}^{3} \alpha_j X_j + \alpha_5 X_5,$$

set $u = u_*$ in the expression for $X\Phi$, and equal zero the result. If the last equality fulfilles identically in x, t only for $\alpha_j = \alpha_5 = 0$, $j = 1, 2, 3$, the desired assertion follows.

By virtue of (2.2)-(2.4), we get

$$\begin{aligned} X\Phi|_{u=u_*} &= -\left[\alpha_1 + \left(\alpha_3 - \frac{2}{3}\alpha_5\right) x\right] \frac{\partial u_*}{\partial x} \\ &- (\alpha_2 + 2\alpha_3 t) \frac{\partial}{\partial x} \left(u_*^{-2/3} \frac{\partial u_*}{\partial x} \right) - 2\alpha_5 u_*. \end{aligned} \tag{2.29}$$

(Here $\partial u_* / \partial t$ is replaced by its expression by virtue of equation (2.9)). Substituting here u_* defined implicitly by formulas (2.21), passing from the derivatives in x to the derivatives in ξ, and then having have recourse to equation (2.20), we transform the invariance condition $X\Phi = 0$ as $u = u_*$ to the form

$$\begin{aligned}
&\left[\alpha_1+\left(\alpha_3-\frac{2}{3}\alpha_5\right)x\right]\left[\xi^4 f^{-5/3}(f+C)-3\xi^5 f^{-2}\right]\\
-&\quad(\alpha_2+2\alpha_3 t)\left[\xi^4 C f^{-5/3}(f+C)+6\xi^4 f^{-2}(f+C)-9\xi^5 f^{-7/3}\right]\\
-&\quad 2\alpha_5\xi^3 f^{-1}=0.
\end{aligned}$$

Let us multiply equality (2.29) by $\xi^{-1}f$ and pass to the limit as $\xi \to 0$. The result of the operation is the relation

$$-(\alpha_2+2\alpha_3 t)\frac{C^4}{9}=0,$$

which has to be satisfied identically in t. If $C \neq 0$, this is possible only as $\alpha_2=\alpha_3=0$. Using the obtained equalities, we have recourse once again to (2.29), multiply this relation by ξ, and then let $\xi \to 0$. Then we obtain the relation

$$2\left[-\alpha_1+\frac{2}{3}\alpha_5\left(-A+\frac{Ct^2}{2}\right)\right]\left(\frac{C}{3}\right)^6=0,$$

whence $\alpha_1=\alpha_5=0$, if $C\neq 0$. Thus, for $C \neq 0$ relation (2.29) is fulfilled identically in ξ and t if $\alpha_j=\alpha_5=0, (j=1,2,3)$, which proves the assertion.

The case $C=0$ is specific. Here the solution of equation (2.20) with the condition $f\to\infty$ as $\xi\to 0$ has the form $f=(3\xi/4)^{3/4}$ and the invariance condition holds if $\alpha_1=\alpha_2=0$, $\alpha_3=2$, $\alpha_5=-9/2$. So, for $C=0$ solution (2.21) renders a solution invariant with respect to the operator

$$5x\partial_x+4t\partial_t-9u\partial_u$$

i.e. a self-similar solution. The parameter representation of this solution is of the following form

$$\begin{aligned}
u_* &= \left(\frac{3}{4}\right)^{-3/4}\xi^3\left(\xi^{-1}-t\right)^{3/4}\\
x &= -4\left(\frac{3}{4}\right)^{3/4}\left(\xi^{-1}-t\right)^{1/4}\left(\frac{9}{5}\xi^{-1}-t\right).
\end{aligned}$$

As shown above, solution (2.21) is not invariant in the ordinary sense, generally speaking. However, it is invariant with respect to a nonlocal transformation addmitted by equation (2.9). This gives the reason to term the solution u_* of this equation its *nonlocal-invariant* solution.

A possibility to employ L-transformation in the search for nonlocal invariant solutions was pointed out yet in [4]. The next example is taken from book [29]. Let us consider equation (2.8). It admits the operator $\tau\partial/\partial\tau - \partial/\partial\omega$. One of the solutions invariant with respect to this operator is

$$\omega = \ln\left[(T-t)^{-1}\frac{\xi^2}{2}\right],$$

where $T > 0$ is a constant. Applying to it L-transformation (1.5), we find a solution of equation (2.7), reciprocical to (2.8),

$$u(x,t) = \left\{\ln\left[\frac{1}{2}(T-t)^{-1}\psi^2(x,t)\right]\right\}. \tag{2.30}$$

Here the function ψ is implicitly defined by the equality

$$x = \psi\left\{\ln\left[(T-t)^{-1}\psi\right] - 1\right\}.$$

One can show that solution (2.30) of equation (2.7) is not invariant with respect to pointwise transformations.

Another possibility of getting nonlocal-invariant solutions is concerned with the consideration of the three-term sequence (1.2), (2.12), (2.14). Papers [13, 17] show a way of finding exact solutions of an equation of this sequence, relying on classical-invariant solutions of the remainders.

In conclusion of the section, let us mention one another method to obtain equations with hidden symmetries. We shall demonstrate it by the example of nonlinear heat-conduction equation (1.2). It is based on the following simple observation: the composition LE of the poitwise equivalence transformation (2.1) and L-transformation conserves the class of equations (1.2), but, generally speaking, $LE \neq EL$.

Consider the equation

$$u_t = \left[\frac{u^m}{(u+1)^{m+2}}u_x\right]_x, \quad m = const, \tag{2.31}$$

$m \neq 0, -2$. Let us apply to equation (2.31) transformation (2.1) with $a = c = 1$, $b = -1$ and then pass in the transformed equation to Lagrangian coordinates. The resulting equation

$$\omega_\tau = [(1-\omega)^m\omega_\xi]_\xi$$

by means of the pointwise transformation $\omega = 1 - \phi$ transformes into the equation

$$\phi_\tau = (\phi^m\phi_\xi)_\xi\,. \tag{2.32}$$

The last equation for $m \neq -4/3$ admits the four-parameter group, while the main group of equation (2.31) with $m \neq -2$ is three-parameter. (If $m = 0$, this equation can be linearized by means of L-transformation). Thus, equation (2.31), equivalent to (2.32) possesses the hidden symmetry property since, evidently, there are no transformations of the type (2.1) transforming one of these equations into another.

The hidden symmetry of equation (2.31) is the source of its non-trivial exact solutions. One of them is constructed in [9]. It can be interpreted as the solution of the one-phase Stefan problem with the heat conductivity coefficient degenerating on the line of the phase transition.

In the family (2.31), one may outline the equation

$$u_t = \left[\frac{u_x}{u^{4/3}(1+u)^{2/3}} \right]_x,$$

equivalent to equation (2.32) with the exeptional value of the exponent $m = -4/3$, [10]. Besides the three-paramater Lie group, it admits a two-parameter group of nonlocal transformations.

3. Lagrange-invariant equations.

Let us term *Lagrange-invariant* (or simply *L-invariant*) the equation that transforms into itself under L-transformation. In this section we confine ourselves to the consideration of L-invariant equations of the type (1.1).

The following problem was first posed in [8]: to find L-invariant equations in the encompassing class of one-dimensional evolution equations of divergent form. This paper was dealing with the class of heat-conduction equations (1.2) which were arbitrary up to a single-variable function $k(u)$. A direct checking shows that equation (1.2) is L-invariant if and only if k satisfies the functional equation

$$k(u) = u^{-2}k(u^{-1}). \tag{3.1}$$

The general solution of this equation has the form $k = u^{-1}\lambda(u)$ where the function λ is subject to the condition $\lambda(u) = \lambda(u^{-1})$ and is arbitrary in all other respects [8].

Note that due to (3.1) L-invariant equation (1.2) is necessarily degenerate at the values $u = 0$, $u = \pm\infty$ or, at least, at one of these values. On the other hand, if we are interested in solutions of equation (3.1) from the class $C^l[a, a^{-1}]$, l is a natural number and $a \in (0,1)$, then we have to set $\lambda \in C^l[a, a^{-1}]$ and, moreover, to claim that all the odd-order derivatives of the function $\mu(v) = \lambda(e^v)$ of the order not exceeding l vanish at the point $v = 0$. It is easy to present examples of such functions: they are

$\lambda_1 = f_1(u)f_1(u^{-1})$, $\lambda_2 = f_2(u) + f_2(u^{-1})$ (here $f_1, f_2 \in C^l[a, a^{-1}]$ are arbitrary), and also an arbitrary function of the arguments $\lambda_1(u)$ and $\lambda_2(u)$. The simplest representative of the class of L-invariant equations (1.2) is the equation

$$u_t = \left(u^{-1}u_x\right)_x, \tag{3.2}$$

arising in the plasma physics.

If equation (1.2) is L-invariant, then in the result of L-transformation each of its solution renders once again a solution to (1.2). The following example shows that the original and the transformed equations may distinguish strongly. Equation (3.2) admits the solution

$$u_1 = \frac{2t}{x^2 + 4t^2}. \tag{3.3}$$

By (1.3), L-transformation of solution (3.3) has the form

$$u_2 = \frac{2t}{(\cos x)^2}. \tag{3.4}$$

These both solutions of equation (3.2) are invariant with respect to some groups of dilatations admitted by this equation. These solutions are substantionally different since there are no Li groups, admitted by equation (3.2) which transfors (3.3) into (3.4). (All such transformations are exhausted by the transfers in x and t, some dilatations in the space x, t, u, and their superpositions). The correspondence between (3.3) and (3.4) is established by the nonlocal L-transformation.

Let us consider now a more general equation of the class (1.1). As was shown in section 1, equation (1.1) is L-invariant if and only if equality (1.6) holds. For the sake of convenience, we present and re-number it once again:

$$\varphi(u,p) = -u\varphi(u^{-1}, -u^{-3}p). \tag{3.5}$$

Relation (3.5) is a functional equation for the function of two variables $\varphi(u,p)$. In what below, we bound ourselves by the search for the solutions of equation (3.5) defined in one of two open half-planes of the plane u, p: $u > 0$ or $u < 0$. A partial justification of this restriction is the fact that given some fixed t, the mapping $\xi = \xi(x,t)$ defined by the first one of formulas (1.3) is diffeomorphic if and only if the smooth function $u(x,t)$ keeps the sign.

So, let in equation (3.5) $u > 0$. Let us make in this equation the substitution

$$\varphi(u,p) = u^{1/2}F(u, u^{-3/2}p). \tag{3.6}$$

Let a continuous function $F(u,p)$ be defined in the half-plane $u>0$ of the plane (u,p), satisfies the equation

$$F(u,q) = -F(u^{-1}, -q) \tag{3.7}$$

and be arbitrary in all other respects. Then the function φ given by (3.6) is a solution of the functional equation (3.6), [9]. The proof of this assertion consists in the direct substitution of (3.6) into (3.5) with the use of property (3.7). Equality (3.7) may be given a more symmetrical form by means of introduction of the function $G(v,q) = F(e^v, q)$; then (3.7) implies the relation

$$G(v,q) = -G(-v,-q) \tag{3.8}$$

(as was observed by H.F.Weinberger).

Thus, the class of L-invariant equations (1.1) is very wide: we can take the function φ arbitrary in the strip $0 < u < 1$, $p \in \mathbf{R}$, and continue it to the half-plane $u > 1$, $p \in \mathbf{R}$ with the help of equalities (3.6), (3.7). (It should be taken into account that the smoothness of the function φ in this strip still does not garantee the smoothness of the continued function at the points of the line $u = 1$). Let us mention in passing that the requested smoothness conditions can be obtained from the representation $\varphi(e^v, e^{3v/2}q) = e^{v/2}G(v,q)$ and relation (3.8)). Let us point out two partial solutions of equation (3.5): the solution linear in p

$$\varphi_1 = u^{-1}f(u)p + u^{1/2}g(u),$$

and the "self-similar" solution, [5],

$$\varphi_2 = u^{-1}p\psi\left(u^{-3/2}p\right).$$

Here $\psi(s)$ is an arbitrary even function; for $0 < u < 1$ the derivatives of the functions f and g satisfy the relations $f(u^{-1}) = f(u)$, $g(u^{-1}) = g(u)$. Besides (3.2), one of the simplest L-invariant equations of the form (1.1) is the equation

$$u_t = \left(|u_x|^{-2/3}u_x\right)_x, \tag{3.9}$$

which appears in considerations of the rectilinear-trajectory motions of non-newtonean liquids (e.g. pseudo-plastic).

Transformation (1.3) defines an automorphism on the set of positive solutions of equations of the type (1.1). One may profit by this property to obtain from an "igniting" solution of equation (1.1) a multi-parameter family of solutions in the case when this equation admits a non-trivial Lie group (wider than the group of transfers in x and t). This method of reproduction of solutions of equation (1.1) is relative to the method of obtaining a family of solutions of an evolution equation by means of differential substitutions developed by S.V.Khabirov [30, 31]. We will demonstrate the idea of this method by the example of equation (3.9).

Equation (3.9) admits the Li group G_5 generated by the operators

$$X_1 = \partial_t, \quad X_2 = \partial_x, \quad X_3 = \partial_u, \quad X_4 = \tfrac{2}{3}t\partial_t + u\partial_u,$$

$$X_5 = \tfrac{2}{3}t\partial_t + x\partial_x - u\partial_u.$$

Let us consider one of the solutions of equation (3.9) invariant with respect to the dilatation transformation with the operator X_5,

$$u_0 = \frac{1}{t^{3/2}}h\left(\frac{x}{t^{3/2}}\right). \tag{3.10}$$

The function $h(\xi)$ satisfies the ordinary differential equation

$$-\frac{3}{2}(\xi h' + h) = \left(|h|^{-2/3}h'\right)'$$

(the prime denotes differentiation in ξ). We are interested in the positive even solutions of the last equation which are regular for all ξ and decrease as $\xi \to \infty$. Such solutions form the one-parameter family

$$h = \left[\frac{27}{16}\xi^4 + \frac{4}{3\sqrt{3}}\left(\frac{I}{C_0}\right)^2\right]^{-1/2}$$

with the parameter C_0. The constant I is defined by the equality

$$I = \int_{\mathbf{R}} \frac{dz}{\sqrt{z^4+1}} \equiv \frac{1}{\sqrt{\pi}}\left[\Gamma\left(\frac{1}{4}\right)\right]^2.$$

This parametrization is convenient because C_0 coincides with the value of the integral which conservs on the solutions of equation (3.9), to name,

$$\int_{\mathbf{R}} u_0(x,t)dx = C_0.$$

The latter makes it possible to interpret (3.10) as a solution of the Cauchy problem

$$u(x,0) = C_0\delta(x)$$

for equation (3.9).

Having subject solution (3.10) to L-transformation (1.3) and taking into account the evenness of $h(\xi)$, we get a new solution of equation (3.9) represented in the form

$$u_1 = t^{3/2}\left\{h\left[H^{-1}(x)\right]\right\}^{-1}, \tag{3.11}$$

where H^{-1} is the inverse to the primitive H of the function h; $H(0) = 0$. Solution (3.11), as well as (3.10), is a self-similar solution (that is, a solution invariant with respect to the dilatation transformation) of equation (3.9), but with another operator X_4. Solution (3.11) possesses some pathological properties: being positive as $t \to 0$, it vanishes as $t = 0$; the function $\varphi(x) = \{h\left[H^{-1}(x)\right]\}^{-1}$ is defined only for $|x| < C_0/2$ and grows unlimitedly as $x \to \pm C_0/2$ so as $\varphi = 2C_0 I^{-1}(C_0/2 \mp x)^{-2} + o(1)$ as $x \to \pm C_0/2$.

Let us now make use of the invariance of (3.9) with respect to the operator X_4. This means that along with the function u_0 the mentioned equation also has as a solution the function $u_0 + C_1$, where C_1 is an arbitrary constant. In what below, we suppose $C_1 > 0$. Let us apply to the function $u_0 + C_1$ L-transformation and denote the result $u_2 = \omega(\xi, t)$. Besides, let us take into account that due to the evenness of the function $h(\xi)$ the second one of the integrals in the right-hand side of the first of relations (1.3) vanishes.

Replacing ξ and x in the final expression (which is possible due to L-invariance of (3.9)), we get the parameter representation for the new solution $u_2(x,t)$ of equation (3.9):

$$x = H\left(\xi/t^{3/2}\right) + C_1\xi, \quad u_2 = t^{3/2}/\left(h\left(\xi/t^{3/2}\right) + C_1 t^{3/2}\right). \tag{3.12}$$

In contrast to solutions (3.10) and (3.11), this solution already depends on two arbitrary constants C_0 and C_1.

The mapping $\mathbf{R} \mapsto \mathbf{R}$ given by the first one of formulas (3.12) for arbitrary fixed t is diffeomorphic for $t > 0$. If $t \to 0+$, this mapping tends to the discontinuous one: $x = C_1\xi + C_0/2$ for $\xi > 0$, $x = C_1\xi - C_0/2$ for $\xi < 0$. The function $u_2(x,t)$ is a classical solution of equation (3.9) for all $t > 0$. Letting $t \to 0$ in the second equality of (3.12) and taking into account that $h(0) > 0$, $h(\xi) = O\left(\xi^{-2}\right)$ as $\xi \to \pm\infty$, we find that this solution satisfies the boundary condition

$$u_2(x,0) = \begin{cases} 1/C_1 & \text{as} \quad x \neq 0, \\ 0 & \text{as} \quad x = 0. \end{cases} \tag{3.13}$$

Thus, started from the "ingniting" self-similar solution (3.10), we have constructed a solution for the Cauchy problem for equation (3.9) with discontinuous initial data (3.13). Solution u_2 not only fails to be self-similar but even is not invariant with respect to any of the Lie groups admitted by equation (3.13). It is curious to remark also that solution (3.12) contains the parameter C_0 which was not involved into the initial conditions. This means non-uniqueness of solution of the Cauchy problem (3.9), (3.13).

Subjecting the solution u_2+C_2 ($C_2 > 0$ is a constant) to transformation (1.3), we get a solution u_3 of equation (3.9), containing three arbitrary parameters. The above-presented method of reproduction of solutions to (3.9) is not the unique possible. The matter is that this equation possesses latent symmetry. One of the nonlocal transformations conserving it is generated by the extension of the Lie group under passing from (3.9) to the equation for the function ω such that $\omega_x = u$. The last equation,

$$\omega_t = |\omega_{xx}|^{-2/3}\omega_{xx},$$

admits the operator $Z = \omega\partial/\partial x$, which is not corresponded by any of the operators of the Lie group G_5, [13, 17]. Another nonlocal transformation is discovered if one passes from equation (3.9) to the equation for the function $v = u_x$. The equation for the function v has the form

$$v_t = \left(|v|^{-2/3}v_x\right)_x$$

and belongs to the number of the hidden-symmetry equations considered in section 2. Combining L-transformation with the above-described nonlocal transformations and the pointwise transformations of the Lie group G_5, we get additional possibilities of multiplicating the number of explicit solutions of equation (3.9).

4. Equations with spherical and cylindrical symmetry

As we have said before, any evolution equation can be interpreted as the mass conservation law for the motion of a fictious liquid. This permits one to perform in it a substitution like (1.3). However, if the original equation contains the space variable explicitly, then, in contrast to (1.4), the transformed equation need not be a differential one. This feature complicates the matter and allows not to perfom in a regular way the search for the nonlocal symmetries in equations of this sort. We present below three partial examples of radially-symmetrical parabolic equations which possess hidden symmetries. Omitting the details, we formulate here only the final results.

The first example is the so-called "the equation of solid H_2", [32]. Seeking for its radially-symmetrical solutions, we arrive to the equation

$$r^{n-1}u_t = \left(r^{n-1}u^{-2}u_r\right)_r. \tag{4.1}$$

were n is the space dimension, $r = (x_1^2 + \dots x_n^2)^{1/2}$. In the case $n = 1$, this equation was studied by M.L.Storm, [33], who applied it to describe the heat transfer in a number of pure metals. It was the paper [33] where the transformation (2.6) linearizing equation (4.1) with $n = 1$ was found.

If $n = 2$, equation (4.1) is equivalent to (2.8), [29]. The last equation admits a four-parameter Lie group while the widest local group of equation (4.1) is three-parameter in this case. Since the connection between (4.1) and (2.8) is nonlocal, this means that equation (4.1) with $n = 2$ has a hidden symmetry.

As for the case $n \geq 3$, with the aid of the composition of both nonlocal and pointwise transformations equation (4.1) can be converted into the porous medium equation (2.3) with $m = 2/(n-2)$ [4, 29]. By the analogy with the previous case we conclude that equation (4.1) for $n \geq 3$ possesses a nonlocal symmetry.

Let us consider now the equation

$$u_t = \Delta \log(u) \tag{4.2}$$

where Δ denotes the Laplace operator in the variables x_1, x_2. This equation occurs in the theory of ultrafine liquid films subject to the van der Waals force, [34]. If the function u does not depend on x_2, equation (4.2) renders (3.2) with $x = x_1$. The last equation belongs to the class of Lagrange-invariant equations.

Another type of invariant solution of (4.2) is $u = v(r,t)$ where $r = (x_1^2 + x_2^2)^{1/2}$ and v is the solution of equation

$$r v_t = \left(r v^{-1} v_r \right)_r . \tag{4.3}$$

J.R.King [35] established that if the function $u(x,t)$ satisfies equation (3.2), then the function

$$v = r^{-2} u(\log(r), t)$$

satisfies equation (4.3). Hence, the latter equation is equivalent to L-invariant one. With the aid of this property a number of exact solutions to equation (4.3) were found in [34, 35, 36].

In conclusion, let us consider the equation

$$r^{n-1} u_t = \left[r^{n-1} |u_r|^l u_r \right]_r . \tag{4.4}$$

where n and l are constants. This equation is a partial case of the one-dimentional parabolic equations "with gradient nonlinearity", [29]. Equation (4.4) outstands from the encompassing equations due to the remarkable property, [36]: if $l \neq -2$ and $l \neq -2(n+1)^{-1}$, there exists a composition of two pointiwise transformations and a nonlocal transformation which transforms (4.4) into an equation of the same type but with n replaced for

$$m = -\frac{(l+2)n}{ln+l+2}.$$

Particularly, if $l = -4(n+2)^{-1}$, then the original and transformed equations coincide. In this case equation (4.4) is like the L-invariant equations studied in section 3. Another interesting case corresponds to the value $l = -2(n+1)(2n+1)^{-1}$. Here $m = 1$, meaning the equivalence of spherically-symmetrical equation (4.4) and the equation

$$v_t = \left(|v_x|^l v_x\right)_x,$$

which describes solutions with plane waves. The latter equation admits a wider (five-parameter) Lie group than (4.4). This means that equation (4.4) with $l = -2(n+1)(2n+1)^{-1}$ has a hidden symmetry.

Let us consider now the exeptional values of the exponent l: $l = -2$ and $l = -2(n+1)^{-1}$. As was shown in [36], in the former case equation (4.4) might be transformed into the linear one, $\omega_t = \omega_{\xi\xi}$. If $l = -2(n+1)^{-1}$, letting

$$r = e^y, \quad u = e^{-ny}\zeta(y,t),$$

we get the following equation for the function ζ:

$$\zeta_t = \left[|\zeta_y - n\zeta|^{-2/(n+1)}(\zeta_y - n\zeta)\right]_y. \tag{4.5}$$

Equation (4.5) does not contain explicitly the new space variable y. In particular, this makes it possible to construct "travelling-wave" solutions to (4.5), $\zeta = f(y - Ct)$, where $C = const$. One of them corresponds to the following exact solution of equation (4.4) with $l = -2(n+1)^{-1}$:

$$u = C^{(n+1)/2} n^{(n-1)/2} \left(r^{2n/(n-1)} + e^{2nCt/(n-1)}\right)^{-(n-1)/2}. \tag{4.6}$$

Unlike the most of explicit (for instance, self-similar) solutions to the degenerate parabolic equations, solution (4.6) is defined for all $t \in \mathbf{R}$. One can interpret it as a solution of "the Cauchy problem with data at $t = -\infty$" for equation (4.4)

$$u \to u_{-\infty}(r) = C^{(n+1)/2} n^{(n-1)/2} r^{-n} \quad \text{as} \quad t \to -\infty.$$

The "initial temperature" $u_{-\infty}(r)$ has a singularity at the point $r = 0$. This function depends on r so that the heat flux Q across the sphere $x_1^2 +$

$\ldots + x_N^2 = r^2$ does not depend on r. The heat flux is equal to the product of the quantity $-|u_r|^{-2/(n-1)} u_r$ and the surface square of this sphere $\Omega_n r^{n-1}$, where Ω_n is the surface square of the unit sphere in $\mathbf{R}^n$. Each value of $Q > 0$ is corresponded by a unique value of the parameter C which determins the family of solutions (4.6),

$$C = n \left(\frac{\Omega_n}{Q} \right)^{2/(n-1)}$$

Acknowledgements

The research presented in this publication was partially supported by Grant n NR6000 from the International Science Foundation. Tho author is very grateful to S.I.Shmarev for the assistance in preparation of the manuscript.

References

1. J.G.Berryman Evolution of a stable profile for a class of nonlinear diffusion equations. III. Slow diffusion on the line. *J.Math.Phys.*, 1980, v.21, pp.1326-1331.
2. A.M.Meirmanov, V.V.Pukhnachov Lagrangian coordinates in the Stefan problem. *Dinamika sploshnoy sredy*, Novosibirsk, 1980, v.47, pp.90-111. (in Russian).
3. M.E.Gurtin, R.C.MacCamy, E.A.Socolovsky A coordinate transformation for the porous media equation that renders the free-boundary stationary. *Quart.Appl.Math.*, 1984, v.42, pp.345-357.
4. V.V.Pukhnachov Evolution equations and coordinates of Lagrange. *Dinamika Sploshnoi Sredi*, Novosibirsk, v.70, 1985, pp.127-141 (in Russian).
5. V.V.Pukhnachov Equivalence transformations and hidden symmetry of evolution equations. *Soviet Math. Dokl.*, 1987, v.35, n.3, pp.555-558; AMS, 1988. Translation of: *Dokladi Akad. Nauk SSSR*, 1987, v.294, n.3, pp.535-538.
6. J.G.Kingston, C.Rogers Reciprocical Böcklund transformations of conservation laws. *Physica Let.*, 1982, v.92 A, n.6, pp.261-264.
7. W.Strampp Böcklund transformations for diffusion equations. *Physica D.*, 1982, n.6, pp.113-118.
8. J.R.Burgan, A.Munier, M.R.Feix, E.Fialkov Homology and the nonlinear heat diffusion equation. *SIAM J. Appl. Math.*, 1984, v.44, n.1, pp.11-18.
9. V.V.Pukhnachov Lagrangian coordinates in free boundary problems for parabolic equations. in *Emerging applications in free boundary problems.*, Proceedings of the International Colloquium *Free Boundary Problems: Theory and Applications*, J.M.Chadam, H.Rasmussen (Editors), Essex: Longman Sci.& Tech., 1993, pp.64-77.
10. L.V.Ovsyannikov The group properties of nonlinear heat conduction equations. *Dokladi Akademii Nauk SSSR*, 1959, v.125, n.3, pp.492-495 (in Russian).
11. L.V.Ovsyannikov Group analysis of differential equations. Moskva, Nauka, 1978 (in Russian). Translation: Academic Press, New York - London, 1982.
12. V.A.Dorodnitsyn, S.R.Svirshchevskii On the Lie-Bäcklund groups admitted by the heat equation with source. Institute of Appl. Math. AN SSSR, Preprint n.101, 1983 (in Russian).
13. I.Sh.Akhatov, R.K.Gazizov, N.H.Ibragimov Quasi-local symmetry of equations of the nonlinear heat conduction. *Soviet Phys. Dokl.*, 1987, v.32, n.7, pp.554-556.

American Institute of Physics, 1988. Translation of: *Dokladi Akad. Nauk SSSR*, 1987, v.295, n.1, pp.75-78 (in Russian).

14. I.Sh.Akhatov, R.K.Gazizov, N.H.Ibragimov Nonlocal symmerties. An euristic approach. in *Itogi nauki i techniki. Seriya: "Sovremennye problemi matematiki. Noveishie dostizjeniya"*. (Series: "The actual problems of mathematics. Newest advances".), v.34, Moskva, VINITI, 1989, pp.3-83 (in Russian).
15. A.M.Vinogradov, I.S.Krasil'shchik A method of computing higher symmetries of nonlinear evolution equations, and nonlocal symmetries. *Soviet Math. Dokl.*, 1980, v.22, n.1, pp.235-239; AMS, 1981. Translation of: *Dokladi Akademii Nauk SSSR*, 1980. v.352, n.6, pp.1289-1293.
16. A.M.Vinogradov, I.S.Krasil'shchik, V.V.Lychagin *Geometry of jet spaces and nonlinear partial differential equations*. Moskva, Nauka, 1986 (in Russian). Translation: New York, London & Beach, 1986.
17. A.Oron, Ph.Rosenau Some symmetries of the nonlinear heat and wave equations. *Phys. Lett. A* , 1986, v.118, n.4, pp.172-176.
18. V.A.Fok The hydrogen atom and non-Euclidean geometry. *Izvestiya Akademii Nauk SSSR. Seria VII. Otdelenie Matematematicheskikh y Estestvennykh nauk*, 1935, n.2, pp.169-179 (in Russian).
19. V.I.Fushchich Additional invariance of relativistic equations. *Theoretical and Mathematical Physics* , 1971, v.7, n.1, pp.323-330; Plenum Publishing Corporation, 1971. Translation of: *Teoret. y Mat. Fisica*, 1971, v.7, n.1, pp.3-12 .
20. V.I.Fushchich A new method of investigating the group properties of the equations of mathematiclal physics. *Soviet Phys. Dokl.*, 1979, v.24, n.6, pp.437-439; American Inst. of Physics, 1980. Translation of: *Dokladi Akademii Nauk SSSR*, 1979, v.246, n.4, pp.846-850.
21. V.I.Fushchich, A.G.Nikitin *Symmetries of Maxwell's equations*. Kiev, Naukova Dumka, 1983 (in Russian). Translation: Dordrecht: D.Reidel Publ., 1987.
22. E.G.Kirnasov On the Walkwist-Eastbruck coverings over heat equations. *Matem. Zametki*, 1987, v.42, n.3, pp.422-434 (in Russian).
23. I.Sh.Akhatov, R.K.Gazizov, N.H.Ibragimov Group classification of the equations of nonlinear filtration. *Soviet Math. Dokl.*, 1987, v.35, n.2, pp.384-386, (AMS 1988). Translation of: *Dokladi Akad Nauk SSSR*, 1987, v.293, n.5, pp.1033-1035 (in Russian).
24. S.R.Svirshchevskii The group properties of the model of heat transfer with relaxation of the heat flux. Inst. Prikladnoy Matematiki AN SSSR. Preprint n.105, 1988, 16 p. (in Russian).
25. B.A.Magadeev On the contact symmetries of nonlinear evolution equations. *Algebra y analis*, 1993, v.5, n.2, pp.141-156 (in Russian).
26. C.Rogers, W.F.Shadwick Böcklund transformations and their applications. New York, Academic Press, Mathematics in Sciences and Engineering Series, 1982.
27. N.H.Ibragimov *Transformation groups applied in mathematical physics*. Moskva, Nauka, 1983 (in Russian). Translated: Dodrecht, D.Reidel Publ., 1985.
28. V.V.Sokolov On the symmetry of evolution equations. *Russian Mathematical Surveys*, 1988, v.43, n.5, pp.165-203; London Math. Society, 1988. Translation of: *Uspehi Mat. Nauk*, 1988, v.43, n.5(263), pp.133-163.
29. A.A.Samarskii, B.A.Galaktionov, S.P.Kurdiumov, A.P.Mikhailov *Blow-up in Problems for Quasilinear Parabolic Equations*. Nauka, Moscow. Translation: Walter de Gruyter, Berlin, to appear.
30. S.V.Khabirov Method of differential substitutions in solution of boundary value problems. In *Differentsial'nye uravneniya s chastnymi proizvodnymi* ("Partial differential equations"), Novosibirsk, Institute matematiki SO AN SSSR, 1984, pp.113-131 (in Russian).
31. S.V.Khabirov Solutions of second-order evolution equations obtained by the differential substitutions. *Dinamika sploshnoy sredi*, Novosibirsk, 1988, v.85, pp.146-161

(in Russian).
32. G.Rosen Nonlinear heat conduction in solid H_2. *Phys. Review B.*, 1979, v.19, third ser., n.4, pp.2398-2399.
33. M.L.Storm Heat conduction in simple metals. *J. Appl. Phys.*, 1951, v.22, pp.940-951.
34. J.R.King. Exact solutions to some nonlinear diffusion equations. *J.Phys. A*, 1990, v.23, pp.3681-3697.
35. O.V.Voinov. Dynamic theory of solid body wetting with a viscous fluid under the action of van der Waals forces. *Zhurnal Prikladnoi Mekhaniki i Tekhnicheskoi Fiziki*, 1994, n.6, pp.69-85 (in Russian).
36. V.V.Pukhnachov. Reciprocal transformations for radial equations of nonlinear heat transfer. *Zapiski Nauchnykh Seminarov POMI*, St.Peterburg, 1994, v.213, pp.151-163 (in Russian).

SPATIAL DECAY ESTIMATES FOR CONE- LIKE SHAPED ELASTIC SOLIDS

RAMÓN QUINTANILLA
Dpt. Matemàtica Aplicada II, U.P.C.
Colom, 11. 08222 Terrassa, Barcelona. Spain.

Abstract. This paper deals with the linear theory of homogeneous elastic solids. We restrict our attention to domains with radial form. For them, we obtain spatial decay estimates of the solutions either for zero traction boundary conditions, for displacement boundary conditions or for other nonlinear boundary conditions. Our main tool is the use of the divergence theorem applied to a vector field which involves a measure of the energy. It is remarkable that this vector field is not usually considered in previous works. Also our method applies to a different class of problems.

1. Introduction

In the last ten years an intensive work on the Saint- Venant principle has been developed. The method used in many papers is the *energy method* and a fundamental tool is the divergence theorem applied to a certain vector field related with the energy. When applied to different problems [1-3, 7-28] such as thermoelasticity [19, 20], viscoelasticity [19] or elasticity with voids [11] the vector fields used are very similar to each other, and the methods exhibit few variations. Good surveys may be found in [4, 5].

Our work has a particular connection with a recent paper of Knops and al. [1] who studied the Saint- Venant principle for linear elastic materials occupying an open region with non- compact boundary. An essential element in their work is that cross- sections are part of a sphere. Also they use the Wirtinger inequality and the Wirtinger' s constants appear in the estimates obtained. The approach is very natural for the zero displacement boundary condition, but the zero boundary traction problem requires a

S. N. Antontsev et al. (eds.), Energy Methods in Continuum Mechanics, 100–111.

postulate on the acting forces over a cross- section. Anyway this postulate is usual in many studies of this problem [4, 5, 9, 13, 15, 18, 22, 24].

In this paper we deal with spatial decay estimates for cone- like regions. Our approach uses an alternative vector field different from those in the previously mentioned references. A function which is different from that in [1] is considered. This function allows easy calculations under many other boundary conditions. The zero boundary traction problem does not require any postulate different from the bounded energy. The displacement boundary problem can be relaxed and the inequalities of the Wirtinger type are not needed. Thus, the estimates do not depend on the geometry of the cross- section, but only on the constitutive coefficients. As a limitation we mention that our method applies only to a subclass of the domains considered at [1]; in fact, we shall restrict our attention to cone- like shaped elastic solids. Our estimates are slower than the ones presented in [1].

In the next Section we state the problems of our work. We also state some equalities and inequalities which are used in later Sections. In Section 3 we prove the main results. Section 4 is devoted to a quantitative calculation in the context of the isotropic and homogeneous elasticity. In Section 5 we study some extensions on the geometry of the domains. We finish in Section 6 by considering some extensions for semi- linear equations.

2. Preliminaries

Throughout this paper we shall employ the usual summation and differentiation conventions: subscripts preceded by a comma denote partial differentiation with respect to the corresponding Cartesian coordinate; $\nabla_{\mathbf{X}}$ is the gradient operator with respect to the place $\mathbf{X}$, summation over repeated subscripts is implied.

We consider an homogeneous linearised elastic material occupying an open region Ω_0 with non- compact boundary and which is contained in the three- dimensional space. Our primary concern is with regions of cone- like form. We suppose that $\Omega_0 = \Omega_1 \cup \Omega_2$ and we select a Cartesian system of coordinates in such a way that

$$0 \in \Omega_1 \subset B(0, r_0) \cap \Omega_0, \tag{1}$$

and

$$\mathbf{X} \cdot \mathbf{N}(\mathbf{X}) = 0, \text{ for all } \mathbf{X} \in \partial\Omega_2, \tag{2}$$

where $B(0, r_0)$ denotes the ball with center at the origin and radius r_0 and $\mathbf{N}(\mathbf{X})$ is the normal vector to $\partial\Omega_2$ at the point $\mathbf{X}$. We introduce some notations for later use. Thus, we denote

$$\Omega(r_1, r_2) = \Omega_0 \cap \Big[B(0, r_1) - B(0, r_2)\Big], \text{ where } r_1 > r_2,$$

and

$$\Sigma(r) = \partial B(0,r) \cap \Omega_0.$$

We recall the equalities

$$|\nabla_{\mathbf{X}} \mathbf{u}|^2 = \left|\frac{\partial \mathbf{u}}{\partial r}\right|^2 + r^{-2}\left|\frac{\partial \mathbf{u}}{\partial \theta}\right|^2 + r^{-2}(\sin^2\theta)^{-1}\left|\frac{\partial \mathbf{u}}{\partial \phi}\right|^2,$$

and

$$\mathbf{X}.\nabla_{\mathbf{X}}\mathbf{u} = r\frac{\partial \mathbf{u}}{\partial r} \text{ for all vector field } \mathbf{u},$$

where r, θ and ϕ are the usual spherical coordinates. It follows that

$$\left|\frac{\partial \mathbf{u}}{\partial r}\right|^2 \leq u_{i,j}u_{i,j}, \text{ for all vector field } \mathbf{u}. \tag{3}$$

We are dealing with the linearised theory of elastostatic, so, the displacement vector field $\mathbf{u}$ satisfies

$$c_{ijkl}u_{k,lj} = 0, \text{ for all } \mathbf{X} \in \Omega_0 - B(0,r_0), \tag{4}$$

and the elasticities satisfy the symmetry condition

$$c_{ijkl} = c_{klij}. \tag{5}$$

The boundary conditions we adjoint to (4) are

$$u_i = f_i, \text{ for all } \mathbf{X} \in \partial\Omega_2, \tag{6}$$

where f_i are given functions such that $\frac{\partial f_i}{\partial r} = 0$ for $i = 1, 2, 3$. Alternatively, we consider zero traction boundary conditions

$$c_{ijkl}u_{k,l}n_j = 0, \text{ for all } \mathbf{X} \in \partial\Omega_2. \tag{7}$$

We recall that non zero boundary conditions have been recently considered in [29]. We also assume that there exists a positive constant C such that

$$c_{ijkl}\psi_{ij}\psi_{kl} \geq C\psi_{ij}\psi_{ij}, \tag{8}$$

for all tensor $\Psi = (\psi_{ij})$. In the next Section, we shall use the equalities

$$(u_i c_{ijkl} u_{k,l})_{,j} = c_{ijkl}u_{i,j}u_{k,l}, \tag{9}$$

and

$$(\frac{1}{2}X_m\delta_{mn}c_{ijkl}u_{i,j}u_{k,l} - X_m c_{injl}u_{j,l}u_{i,m} + u_i c_{injl}u_{j,l})_{,n} = \frac{3}{2}c_{ijkl}u_{i,j}u_{k,l}, \tag{10}$$

where δ_{mn} is the Kronecker delta and X_m are the components of $\mathbf{X}$. Thus, we deduce

$$(\frac{1}{2}X_m\delta_{mn}c_{ijkl}u_{i,j}u_{k,l} - X_m c_{injl}u_{j,l}u_{i,m})_{,n} = \frac{1}{2}c_{ijkl}u_{i,j}u_{k,l}. \qquad (11)$$

In previous works equality (9) is usually considered as a departure point to use the energy methods. Equality (10) is a particular case of one established by Green [31] who stated it in the nonlinear theory. An alternative approach of this equality (also in the nonlinear case) can be found in the book by Olver [6] (p.283). It is also remarkable that Knops and Stuart [30] made use of this equality to deduce uniqueness of solutions for a displacement boundary value problem in the context of nonlinear elasticity. Recently, Quintanilla [32] has extended some of this results to mixtures of elastic solids.

3. Estimates in linearised theory

In this Section we obtain decay estimates for the boundary value problem determined by the equations (3) and the boundary conditions (6) or (7), or the mixed boundary conditions).

To this end we introduce the function

$$W(r) = r\int_{\Sigma(r)} P ds, \qquad (12)$$

where

$$P = c_{ijkl}(\frac{1}{2}u_{i,j} - \frac{\partial u_i}{\partial r}\frac{X_j}{r})u_{k,l}, \qquad (13)$$

In view of the divergence theorem and the geometric condition (2) we may write

$$W(r) = \frac{1}{2}\int_{\Omega(r,r_0)} c_{ijkl}u_{i,j}u_{k,l}dv + W(r_0), \text{ for all } r \geq r_0, \qquad (14)$$

which is valid for the displacement boundary conditions (6), for zero traction boundary conditions (7) or for mixed boundary conditions. Direct differentiation gives

$$W'(r) = \frac{1}{2}\int_{\Sigma(r)} c_{ijkl}u_{i,j}u_{k,l}ds. \qquad (15)$$

From (8), we clearly have

$$W'(r) \geq 0, \text{ for all } r \geq r_0.$$

Now, we derive an inequality which is essential in our approach. From the Schwarz inequality, we have

$$
\begin{aligned}
&|W(r)| \\
&\leq r\left|\int_{\Sigma(r)} c_{ijkl}u_{i,j}u_{k,l}ds\right|^{\frac{1}{2}} \\
&\qquad \left|\int_{\Sigma(r)} c_{ijkl}(\frac{1}{2}u_{i,j}-\frac{\partial u_i}{\partial r}\frac{X_j}{r})(\frac{1}{2}u_{k,l}-\frac{\partial u_k}{\partial r}\frac{X_l}{r})ds\right|^{\frac{1}{2}} \\
&= r\left|\int_{\Sigma(r)} c_{ijkl}u_{i,j}u_{k,l}ds\right|^{\frac{1}{2}} \\
&\qquad \left|\int_{\Sigma(r)} c_{ijkl}[\frac{1}{4}u_{i,j}u_{k,l}+\frac{\partial u_i}{\partial r}\frac{X_j}{r}\frac{\partial u_k}{\partial r}\frac{X_l}{r}-u_{i,j}\frac{\partial u_k}{\partial r}\frac{X_l}{r}]ds\right|^{\frac{1}{2}} \\
&\leq k^{\frac{1}{2}}rW'(r),
\end{aligned}
\tag{16}
$$

where k is such that

$$
\left|c_{ijkl}\left[\frac{1}{4}u_{i,j}u_{k,l}+\frac{\partial u_i}{\partial r}\frac{X_j}{r}\frac{\partial u_k}{\partial r}\frac{X_l}{r}-u_{i,j}\frac{\partial u_k}{\partial r}\frac{X_l}{r}\right]\right| \leq \frac{k}{4}c_{ijkl}u_{i,j}u_{k,l}. \tag{17}
$$

We may take

$$
\frac{k}{4}=\frac{1}{4}+\frac{12M}{C},
$$

with $M=\sup|c_{ijkl}|$ and C is given at (8).

Now, if we suppose that $W(r_0)>0$, we may conclude

$$
W(r)\geq W(r_0)(\frac{r}{r_0})^{\gamma}, \text{ for all } r\geq r_0, \tag{18}
$$

where $\gamma=k^{-\frac{1}{2}}$. This implies

$$
\lim_{r\to\infty} W(r)=\infty.(\text{ algebraicly})
$$

The energy function

$$
E(r)=\frac{1}{2}\int_{\Omega(\infty,r)} c_{ijkl}u_{i,j}u_{k,l}dv, \tag{19}
$$

does not exist when the fuction $W(r)$ is positive at any point.

Thus, if we postulate the existence of the energy contained in Ω_0, then $W(r) \leq 0$ for all r. An integration provides the estimate

$$-W(r) \leq -W(r_0)(\frac{r}{r_0})^{-\gamma}, \text{ for all } r \geq r_0. \tag{20}$$

In case $W(r_0) = 0$ it follows that $W(r) = 0$ for all $r \geq r_0$ and we conclude

$$u_i \equiv a_i, \text{ for all } r \geq r_0.$$

If $W(r_0) < 0$, we obtain the following estimate for the energy

$$E(r) \leq E(r_0)(\frac{r}{r_0})^{-\gamma}, \text{ for all } r \geq r_0. \tag{21}$$

We have proved

Theorem 1. For the boundary conditions (6) or (7), or mixed boundary conditions), the energy function $E(r)$ defined at (19) does not exist, vanishes identically or decays algebraicly for all r.

Remark. Our analysis may be adapted to study problems in exterior regions to some bounded domain Ω^* in a way similar to that followed in [1]. We take

$$W(r) = r \int_{\partial B(0,r)} P ds,$$

where P was defined at (13). We may obtain spatial decay for the energy (19) whenever it is bounded. Note that our approach does not need the Wirtinger inequality and normalisation, neither any self- equilibrating condition needed in [1] (p. 328).

Remark. Uniqueness theorems can be also obtained in a similar way than in [1].

Remark. In order to show the versatility of our approach we are going to study the problem determined by the alternative boundary conditions. We relax (7) to

$$c_{ijkl}u_{k,l}n_j + g_i(\mathbf{u}) = 0, \text{ for all } \mathbf{X} \in \partial\Omega_2, \tag{22}$$

where

$$g_i = \frac{\partial F}{\partial u_i}(\mathbf{u}), F(0) = 0, \text{ and } F \geq 0. \tag{23}$$

Conditions of this kind have been studied recently by Horgan and Payne [3] for the Laplace equation in cylinders of constant section. Their studies are related to problems of boundary conditions in the context of heat transfer.

The linear problem for the boundary conditions can be stated as

$$c_{ijkl}u_{k,l}n_j + m_{ih}u_h = 0, \text{ for all } \mathbf{X} \in \partial\Omega_2,$$

where $M = (m_{ij})$ is a semi- definite symmetric matrix.

In the general case, we may define

$$W_F(r) = W(r) + r\int_{\partial\Sigma(r)} F(\mathbf{u})dl.$$

In this case the divergence theorem gives

$$W(r) + \int_{\partial\Omega(r,r_0)} r\frac{\partial u_i}{\partial r}g_i(\mathbf{u})ds = \frac{1}{2}\int_{\Omega(r,r_0)} c_{ijkl}u_{i,j}u_{k,l}dv + W(r_0), \tag{24}$$

for all $r \geq r_0$.

Direct differentiation gives

$$W'(r) = \frac{1}{2}\int_{\Sigma(r)} c_{ijkl}u_{i,j}u_{k,l}ds - r\int_{\partial\Sigma(r)} \frac{\partial F}{\partial r}(\mathbf{u})dl. \tag{25}$$

and

$$W_F'(r) = \frac{1}{2}\int_{\Sigma(r)} c_{ijkl}u_{i,j}u_{k,l}ds + 2\int_{\partial\Sigma(r)} F(\mathbf{u})dl.$$

If we define

$$E_F(r) = E(r) + 2\int_{\partial\Omega(\infty,r)} F(\mathbf{u})ds,$$

and we suppose that E and E_F are defined, we may obtain an spatial decay similar to (21)

$$E_F(r) \leq E_F(r_0)(\frac{r}{r_0})^{-\gamma}, \tag{26}$$

where $\gamma = k^{-\frac{1}{2}}$.

This method also works in case that the boundary conditions (22) are combined with (6) and, or (7).

4. Isotropic linear theory

It has been previously commented that our approach does not use the Wirtinger inequality and it allows to obtain rates for the decay estimate which do not depend on the geometry, but only on the constitutive coefficients. Previous estimates are not clear because of the cumbersome calculations which the linearised case involve. For isotropic and homogeneous elastic bodies the calculations are easier and the dependence on the constitutive coefficients is more transparent. In this Section, we only consider boundary conditions of type (6).

First we recall that for a linear isotropic and homogeneous elastic solid the equilibrium equations are given by

$$u_{i,jj} + \nu u_{j,ji} = 0, \ \ \mathbf{X} \in \Omega_0 - B(0, r_0), \ \text{for } (\frac{1}{3} < \nu < \infty) \tag{27}$$

where $\nu = (1-2\sigma)^{-1}$ and σ is the Poisson ratio. As in the previous Section, we introduce

$$I(r) = r \int_{\Sigma(r)} Q ds, \tag{28}$$

where

$$Q = (u_{i,j} + \delta_{ij} \nu u_{k,k})(\frac{1}{2} u_{i,j} - \frac{\partial u_i}{\partial r} \frac{X_j}{r}).$$

From the divergence theorem we derive a quantity which is related with a "quasi- energy"

$$I(r) = \frac{1}{2} \int_{\Omega(r,r_0)} [u_{i,j} u_{i,j} + \nu u_{i,i} u_{j,j}] dv + I(r_0). \tag{29}$$

Direct differentiation gives a "quasi- energy"

$$I'(r) = \frac{1}{2} \int_{\Sigma(r)} [u_{i,j} u_{i,j} + \nu u_{i,i} u_{j,j}] ds, \tag{30}$$

Now, we have

$$\frac{1}{r} I(r) = \frac{1}{2} \int_{\Sigma(r)} [u_{i,j} u_{i,j} + \nu u_{i,i} u_{j,j}] ds - \int_{\Sigma(r)} [\frac{\partial u_i}{\partial r} \frac{\partial u_i}{\partial r} + \nu u_{i,i} \frac{\partial u_j}{\partial r} \frac{X_j}{r}] ds,$$

An easy calculation shows that

$$|u_{i,i} \frac{\partial u_j}{\partial r} \frac{X_j}{r}| \leq (u_{i,i} u_{j,j})^{\frac{1}{2}} (\frac{\partial u_j}{\partial r} \frac{\partial u_j}{\partial r})^{\frac{1}{2}}, \tag{31}$$

and

$$\frac{1}{r} |I(r)| \leq \frac{1}{2} | \int_{\Sigma(r)} (u_{i,j} u_{i,j} - 2 \frac{\partial u_j}{\partial r} \frac{\partial u_j}{\partial r}) ds|$$

$$+ \nu (\frac{\epsilon + 1}{2} \int_{\Sigma(r)} u_{i,i} u_{j,j} ds + \frac{1}{2\epsilon} \int_{\Sigma(r)} \frac{\partial u_j}{\partial r} \frac{\partial u_j}{\partial r} ds).$$

Taking $\epsilon = \nu^{\frac{1}{2}}$, we deduce

$$\frac{1}{r} |I(r)| \leq (1 + \nu^{\frac{1}{2}}) I'(r). \tag{32}$$

From the previous inequality, it follows that $I(r)$ grows algebraicly as r^γ, where

$$\gamma = (\nu^{\frac{1}{2}} + 1)^{-1},$$

or the quasi- energy

$$E^*(r) = \frac{1}{2}\int_{\Omega(\infty,r)} [u_{i,j}u_{i,j} + \nu u_{i,i}u_{j,j}]dv,$$

decays algebraicly as $r^{-\gamma}$.

We recall that the decay estimate obtained in [1] for the isotropic problem is

$$\gamma^* = \frac{2}{\sqrt{k(1+\nu)}},$$

where k is the Wirtinger constant.

Remark. The remark stated in the previous Section for exterior domains is also valid for isotropic and homogeneous elastic solids. Thus, we may obtain decay estimates for the "quasi- energy" contained in the three-dimensional domains whenever they are bounded.

5. Extension to other geometries

As in the previous sections we suppose that $\Omega_0 = \Omega_1 \cup \Omega_2$. We select a Cartesian system of coordinates in such a way that

$$0 \in \Omega_1 \subset B(0, r_0) \cap \Omega_0. \tag{33}$$

Now, we relax condition (2) to

$$\mathbf{X} \cdot \mathbf{N}(\mathbf{X}) \leq 0, \text{ for all } \mathbf{X} \in \partial\Omega_2. \tag{34}$$

If we relax the geometry of the domains where our methods can be used, we have to restrict our boundary conditions to the case (7).

If we define $W(r)$ by the equality (12), we deduce

$$W(r)$$
$$= \frac{1}{2}\int_{\Omega(r,r_0)} c_{ijkl}u_{i,j}u_{k,l}dv - \frac{1}{2}\int_{\partial\Omega(r,r_0)} \mathbf{X} \cdot \mathbf{N}(\mathbf{X})c_{ijkl}u_{i,j}u_{k,l}ds + W(r_0), \tag{35}$$

for all $r \geq r_0$. Direct differentiation gives

$$W'(r) = \frac{1}{2}\int_{\Sigma(r)} c_{ijkl}u_{i,j}u_{k,l}ds - \frac{1}{2}\int_{\partial\Sigma(r)} \mathbf{X} \cdot \mathbf{N}(\mathbf{X})c_{ijkl}u_{i,j}u_{k,l}dl, \tag{36}$$

for all $r \geq r_0$.

Now, we define

$$F_\partial = \frac{1}{2}\int_{\partial\Omega(\infty,r)} \mathbf{X}\cdot\mathbf{N}(\mathbf{X})c_{ijkl}u_{i,j}u_{k,l}ds.$$

In a similar way that in Section 3 we have that $E - F_\partial$ does not exists when the function $W(r)$ is positive at some point. An integration provides the estimate

$$E(r) - F_\partial(r) \leq (E(r_0) - F_\partial(r_0))(\frac{r}{r_0})^{-\gamma}, \tag{37}$$

where γ was determined in Section 3 whenever the energy and F_∂ are bounded.

It is also possible to consider boundary conditions of type (22).

6. Semi-linear systems

In this Section we extend the approach introduced in the previous Section to a class of semi- linear systems. Spatial decay estimates for this kind of systems were studied in [17, 33, 34].

We consider the system

$$c_{ijkl}u_{k,lj} = h_i(u), \text{ for all } \mathbf{X} \in \Omega_0 - B(0,r_0), \tag{38}$$

where

$$h_i = \frac{\partial H}{\partial u_i}(\mathbf{u}), \; H(0) = 0, \text{ and } H \geq 0. \tag{39}$$

In this case we may derive the equality

$$\left(X_m\delta_{mn}(\frac{1}{2}c_{ijkl}u_{i,j}u_{k,l} + H(\mathbf{u})) - X_m c_{injl}u_{j,l}u_{i,m}\right)_{,n}$$

$$= \frac{1}{2}c_{ijkl}u_{i,j}u_{k,l} + 3H(\mathbf{u}). \tag{40}$$

We define

$$W_H(r) = W(r) + r\int_{\Sigma(r)} H(\mathbf{u})dl,$$

and the divergence theorem and the geometry of the domain lead to

$$W_H(r) = \frac{1}{2}\int_{\Omega(r,r_0)} (c_{ijkl}u_{i,j}u_{k,l} + 6H(\mathbf{u}))dv + W_H(r_0), \text{ for all } r \geq r_0. \tag{41}$$

Direct differentiation gives

$$W_H'(r) = \frac{1}{2}\int_{\Sigma(r)} (c_{ijkl}u_{i,j}u_{k,l} + 6H(\mathbf{u}))ds.$$

Similar results to those obtained in the previous Sections can be obtained for the energy. That is

$$E_H(r) = E(r) + 3\int_{\Omega(\infty,r)} H(\mathbf{u})dv.$$

It is also possible to consider boundary conditions of type (22).

Remark. As in the case $h_i = 0$, we may extend our analysis to exterior domains. A special case is the whole space. Now, $W_H(r) \geq 0$ for all $r > 0$. When the total energy is bounded , we must have $W_H(r) = 0$ for all $r > 0$. It follows that $\mathbf{u}$ is constant and satisfies $H(\mathbf{u}) = 0$.

References

1. R.J.KNOPS, S. RIONERO & L.E. PAYNE. *Saint- Venant's principle on unbounded regions.* Proc. Roy. Soc. Edinburgh, 115A:319–336, 1990.
2. R.QUINTANILLA. *A spatial decay estimates for the hyperbolic heat equation.* Accepted for publication in S.I.A.M. Jour. Math. Anal.
3. C.O. HORGAN & L. E. PAYNE. *Phragmen- Lindeloff type results for harmonic functions with nonlinear boundary conditions.* Arch. Rational Mech. Anal., 122:123–144, 1993.
4. C.O. HORGAN & J.K. KNOWLES. *Recent developments concerning Saint-Venant' s Principle,* in "Advances in Applied Mechanics" (Ed. J.W. Hutchinson & T.Y. Wu), 23:179–269, Academic Press, New York, 1983.
5. C.O. HORGAN. *Recent developments concerning Saint-Venant' s Principle: An update.* Applied Mechanics Reviews, 42:295–303, 1989,
6. P. J. OLVER. "Aplications of Lie Groups to Differential Equations". Springer-Verlag, Berlin, 1986.
7. J.N. FLAVIN & R.J. KNOPS. *Some spatial decay estimates in continuum dynamics.* Jour. Elasticity, 17:249–264, 1987.
8. J.N. FLAVIN, R.J. KNOPS & L.E. PAYNE. *Some decay estimates for the constrained elastic cylinder of variable cross section.* Quat. Appl. Mathematics, XLVII:325–350, 1989.
9. C.O. HORGAN & L.E. PAYNE. *Decay estimates for second order quasilinear partial differential equations transient heat conduction.* Adv. Appl. Mathematics, 5:309–332, 1984.
10. R. QUINTANILLA. *Some theorems of Phragmen- Lindeloff type for nonlinear differential equations.* Publicacions Matematiques, 37:443– 463, 1993.
11. D.IESAN & R. QUINTANILLA. *Decay estimates and energy bounds for porous elastic cylinders.* Accepted in Jour. Appl. Math. Phys.(ZAMP).
12. C.O.HORGAN & L.E.PAYNE. *Decay Estimates for a Class of Second-Order Quasilinear Equations in three Dimensions.* Arch. Rat. Mech. Anal., 86:279–289, 1984.
13. R.J.KNOPS & L.E.PAYNE. *A Saint-Venant Principle for Nonlinear Elasticity .* Arch. Rat. Mech. Anal., 81:1–12, 1983.
14. C.O.HORGAN & L.E.PAYNE. *Decay Estimates for a class of Nonlinear Boundary Value Problems in two Dimensions.* S.I.A.M. Jour. Math. Anal. , 20:782–788, 1989.
15. C.O.HORGAN & L.E.PAYNE. *On Saint-Venant's Principle in finite Anti- Plane Shear: An Energy Approach.* Arch. Rat. Mech. Anal., 109:107–137, 1990.
16. J.N.FLAVIN, R.J.KNOPS & L.E.PAYNE. *Decay Estimates for Constrained Elastic Cylinder of Variable Cross Section.* Quat. Appl. Mathematics , XLVII:325–350, 1989.

17. J.N.FLAVIN, R.J.KNOPS & L.E.PAYNE. *Asymptotic behavior of solution to semilinear elliptic equations on the half-cylinder.* Jour. Appl. Math. Phys.(ZAMP), 43:405–421, 1992.
18. R. A. TOUPIN. *Saint- Venant' s principle.* Arch. Rat. Mech. Anal., 18:83–96, 1965.
19. R.J. KNOPS & C.LUPOLI. *Some recent results on Saint- Venant's principle.* To appear.
20. C. LUPOLI. *A Phragmèn- Lindeloff principle for the thermoelastic cylinder of variable cross- section.* Accepted in Meccanica.
21. J. N. FLAVIN, R. J. KNOPS & L. E. PAYNE. *Energy bounds in dynamical problems for a semi- infinite elastic beam,* in "Elasticity : Mathematical Methods and Applications" (G. EASON and R. W. OGDEN, Eds.), 101–111, Ellis- Horwood, Chichester, 1989.
22. R.J. KNOPS. *A Phragmen-Lindeloff theorem for the free elastic cylinder.* Rendiconti di Matematica, Serie VII, 10:601–622, 1990.
23. G.P. GALDI, R. J. KNOPS & S. RIONERO. *Asymptotic behaviour in the nonlinear elastic beam.* Arch. Rat. Mech. Anal., 87:305–318, 1985.
24. N. WECK. *A explicit Saint Venant' s principle in three dimensional elasticity* in "Ordinary and Partial Differential Equations, Dundee(1976)". Lecture Notes in Mathematics, 564:518–526, Springer, 1976.
25. J. J. ROSEMAN & S.ZIMERING. *On the spatial decay of the energy for some quasilinear boundary value problems in semi- infinite cylinders.* Jour. Math. Anal. Appl., 139:194–204, 1989.
26. Y. SHENKER & J. J. ROSEMAN. *On the Saint Venant' s principle in the case of infinite energy.* J. Elasticity, 30:55–68, 1993.
27. S. BREUER & J. J. ROSEMAN. *Spatial decay results for a class of quasilinear elliptic equations of mode zero.* Diff. Int. Equations, 4:411–423, 1991.
28. S. BREUER & J. J. ROSEMAN. *On the spatial energy decay boundary value problems in cone- like and exterior domains.* Diff. Int. Equations, 2:310–325, 1989.
29. J. N. FLAVIN & S. RIONERO. *Decay and other estimates for an elastic cylinder.* Quat. Jour. Mech. Appl. Math., 46:299–309, 1993.
30. R. J. KNOPS & C. A. STUART. *Quasiconvexity and uniqueness of equilibrium solutions in nonlinear elasticity.* Arch. Rational Mech. Anal., 86:233–149, 1984.
31. A. E. GREEN. *On some general formulae in finite elastostatics.* Arch. Rat. Mech. Anal., 50:73–80, 1973.
32. R. QUINTANILLA. *Uniqueness of equilibrium solutions in nonlinear theory of elastic mixtures.* Accepted in ZAMM.
33. J.N.FLAVIN, R.J.KNOPS & L.E.PAYNE. *Asymptotic and other estimates for a semilinear elliptic equation in a cylinder.* Quat. Jour. Mech. Appl. Math. , 45:617–627, 1992.
34. J.I. DIAZ & O.A. OLEINICK. *Nonlinear elliptic boundary-value problems in unbounded domains and the asymptotic behaviour of its solutions.* C. R. Acad. Sci. Paris, 315:787–792, 1992.

ENERGY FLUID MOTIONS STABILITY FOR FREE BOUNDARY LIKE PROBLEMS IN THE EXTERIOR OF CONVEX STARSHAPED DOMAINS

S. RIONERO
Dipartimento di Matematica e Applicazioni "R. Caccioppoli"
Universitá degli Studi di Napoli "Federico II"
Complesso Monte S. Angelo – Edificio "T" – Via Cintia
80126 Napoli (Italy)

Abstract - The Energy stability of viscous incompressible fluid motions occurring in the exterior of convex starshaped domains under mixed boundary conditions is considered. Stability criteria are obtained.

1. Introduction

Let $\Omega \subset I\!R^3$ be a region filled by a viscous incompressible isothermal newtonian fluid and let $\partial\Omega = \Sigma \cup \Gamma$, $\Sigma \cap \Gamma = \emptyset$ be smooth with Σ rigid and Γ "free but invariable" (i.e. a system of coordinates can be chosen on which it does not change). On Σ we assume the adherence conditions for the velocity field while on Γ we assume the usual slip conditions i.e. we assign the normal component of the velocity and the tangential component of the stress vector. This mixed boundary value problem belongs to the class of the free boundary like problems because represents one of the first step toward more general problems of the motion of a fluid in a domain with free boundary. When Ω is bounded at least in one direction, this problem has been studied by several authors (see e.g. [1-9] and the references in [2]). When Ω is unbounded in all directions this problem has been deeply studied in the case of the homogeneous boundary conditions [12] but, as far as we know, only few results have been obtained in the case of the mixed boundary data. In

S. N. Antontsev et al. (eds.), Energy Methods in Continuum Mechanics, 112–125.

particular, the exterior of a sphere has been recently considered in [13], and criteria of energy stability have been obtained. Here we consider the fluid motions in the exterior of convex starshaped domains and after having obtained an L^2- energy equality (sect.2), in section 3 we obtain some suitable weighted inequalities of the Sobolev type. Finally in section 4 we give some criteria of stability.

2. Statement of the problem

Let Ω be an unbounded domain filled by a viscous incompressible isothermal fluid with smooth boundary $\partial\Omega = \Sigma \cup \Gamma$, where Σ is assumed "rigid" and Γ is the "free invariable" boundary[1]. As equations of motion we consider the Navier-Stokes equations with $T =$ const. > 0:

$$\begin{cases} \mathbf{v}_t + \mathbf{v}\cdot\nabla\mathbf{v} = -\nabla p + \nu\Delta\mathbf{v} + \mathbf{F} \\ \qquad\qquad\qquad\qquad\qquad\qquad (\mathbf{x},t) \in \Omega \times [0,T] \\ \nabla\cdot\mathbf{v} = 0 \end{cases} \tag{2.1}$$

where $\mathbf{v}$ represents the velocity field, p the pressure field divided by the constant density, $\mathbf{F}(\mathbf{x},t)$ the body force and $\nu(>0)$ is the kinematic viscosity. To equation (2.1) we append the initial conditions

$$\mathbf{v}(\mathbf{x},t) = \mathbf{v}_0(\mathbf{x}) \tag{2.2}$$

where $\mathbf{v}_0$ is a divergence free assigned velocity field and the mixed boundary conditions

$$\begin{cases} \mathbf{v} = \mathbf{a}(\mathbf{x},t) & \text{on } \Sigma \\ \mathbf{v}\cdot\mathbf{n} = \mathbf{a}_1(\mathbf{x},t), \quad \mathbf{D}(\mathbf{v})\mathbf{n} - (\mathbf{n}\cdot\mathbf{D}(\mathbf{v})\cdot\mathbf{n})\mathbf{n} = \mathbf{b} & \text{on } \Gamma \end{cases} \tag{2.3}$$

where $\mathbf{a}$, $\mathbf{a}_1$, $\mathbf{b}$ (with $\mathbf{b}\cdot\mathbf{n} = 0$) are assigned vector fields, $\mathbf{n}$ is the unit outer normal to $\partial\Omega$ and $\mathbf{D}(\mathbf{v}) = 1/2\{\nabla\mathbf{v} + (\nabla\mathbf{v})^T\}$ is the rate of deformation tensor. $(2.3)_2$ and $(2.3)_3$ represent the slip condition. In particular $(2.3)_3$ corresponds to assigne the tangential component of the stress vector $\mathbf{t} = -p\mathbf{n} + 2\nu\mathbf{D}(\mathbf{v})\cdot\mathbf{n}$.

If $(\mathbf{v}+\mathbf{u}, p+\pi)$ is the motion corresponding to the perturbation $\mathbf{u}_0 = \mathbf{u}_0(\mathbf{x})$ in the initial data, then the perturbation $(\mathbf{u},\pi)$ satisfies the following initial

[1] Either Σ or Γ may be empty.

boundary value problem

$$\begin{cases} \mathbf{u}_t + (\mathbf{v}+\mathbf{u})\cdot\nabla\mathbf{u} + \mathbf{u}\cdot\nabla\mathbf{v} = -\nabla\pi + \nu\Delta\mathbf{u} \\ & (\mathbf{x},t)\in\Omega\times[0,T] \\ \nabla\cdot\mathbf{u} = 0 \end{cases} \tag{2.4}$$

$$\begin{cases} \mathbf{u}(\mathbf{x},0) = \mathbf{u}_0(\mathbf{x}) & \text{on } \Omega \\ \mathbf{u} = 0 & \text{on } \Sigma \\ \mathbf{u}\cdot\mathbf{n} = \mathbf{n}\cdot\mathbf{d}\times\mathbf{n} = 0 & \text{on } \Gamma\,. \end{cases} \tag{2.5}$$

In the sequel we shall be concerned with solutions of the above problem (2.4)–(2.5) which satisfy the following energy equality

$$\begin{aligned} \frac{d}{dt}\int_\Omega f u^2 d\Omega =& 2\int_\Omega [f\mathbf{u}\cdot\nabla\mathbf{u}\cdot\mathbf{v} + (\nabla f\cdot\mathbf{u})\mathbf{u}\cdot\mathbf{v} + \pi\nabla f\cdot\mathbf{u} + \frac{\nu}{2}u^2\Delta f]d\Omega + \\ & -2\int_\Omega f[(\mathbf{u}+\mathbf{v})\cdot\nabla\mathbf{u}\cdot\mathbf{u} + \nu\nabla\mathbf{u} : \nabla\mathbf{u}]d\Omega + \\ & +\nu\int_\Gamma [2f(\mathbf{u}\cdot\nabla\mathbf{n}\cdot\mathbf{u} - \mathbf{u}\cdot\nabla u_n - u^2\frac{\partial f}{\partial\mathbf{n}}]d\Gamma \end{aligned} \tag{2.6}$$

where $f(\mathbf{x})$ is any weight function. Of course to this class of solutions belong those solutions to (2.4)–(2.5) whose behaviour at large spatial distances is suitably linked to that of f. Formally, the equality (2.6) can be obtained by integrating (2.4), multiplied by $f\mathbf{u}$ and taking into account the following identities for divergence free vectors $\mathbf{u}$ verifying $(2.5)_{3,4}$

i) $f\mathbf{u}\cdot\Delta\mathbf{u} = \nabla\cdot(f\nabla\mathbf{u}\cdot\mathbf{u} - \frac{1}{2}u^2\nabla f) + \frac{1}{2}u^2\Delta f - f\nabla\mathbf{u} : \nabla\mathbf{u}$

ii) $f\mathbf{u}\cdot\nabla\mathbf{v}\cdot\mathbf{u} = \nabla\cdot[f(\mathbf{u}\cdot\mathbf{v})\mathbf{u}] - f\mathbf{u}\cdot\nabla\mathbf{u}\cdot\mathbf{v} - (\nabla f\cdot\mathbf{u})\mathbf{u}\cdot\mathbf{v}$

iii) $f\nabla\pi\cdot\mathbf{u} = \nabla\cdot(f\pi\mathbf{u}) - \pi\nabla f\cdot\mathbf{u}$

iv) $\mathbf{n}\cdot\nabla\mathbf{u}\cdot\mathbf{u} = 2\mathbf{n}\cdot\mathbf{d}\cdot\mathbf{u} - \mathbf{u}\cdot\nabla\mathbf{u}\cdot\mathbf{n} = -\mathbf{u}\cdot\nabla\mathbf{u}\cdot\mathbf{n} =$
$= -\mathbf{u}\cdot\nabla u_n + \mathbf{u}\cdot\nabla\mathbf{n}\cdot\mathbf{u}$

Remark 1 - In the sequel we will consider the sole case in which Γ is constituted by parts Γ_i which belong to plane, spherical or cylindrical surfaces. Then it is an easy matter to show that it turns out

a) $\mathbf{u}\cdot\nabla u_n = 0$, $\mathbf{u}\cdot\nabla\mathbf{n}\cdot\mathbf{u} = 0$ on a plane surface

b) $\mathbf{u}\cdot\nabla u_n = 0$, $\mathbf{u}\cdot\nabla\mathbf{n}\cdot\mathbf{u} = \pm\frac{u_\theta^2}{\rho_0}$ on a cylindrical surface where ρ_0 is the radius (holding always the same sign)

c) $\mathbf{u}\cdot\nabla u_n = 0$, $\mathbf{u}\cdot\nabla\mathbf{n}\cdot\mathbf{u} = \pm\frac{u^2}{\rho_0}$ on a spherical surface, where ρ_0 is the radius (holding always the same sign).

In fact in the case a), let $z = 0$ be the equation of Γ_i. Then from $u_n = u_z \equiv 0$ on Γ_i it follows that $\frac{\partial u_z}{\partial x} = \frac{\partial u_z}{\partial y} \equiv 0$ and hence $\mathbf{u}\cdot\nabla u_n \equiv 0$ on Γ_i. Further, because $\mathbf{n} = \nabla z$ is constant, $\nabla\mathbf{n} = 0$. In the case b) let $\rho = \rho_0 = \text{const.} > 0$ be the equation of Γ_i in a cylindrical frame of reference (O, ρ, θ, z) and let $\mathbf{e}_\rho = \nabla\rho$. It follows that $\mathbf{n} = \pm\mathbf{e}_\rho$. From $u_n \equiv 0$ on Γ_i it turns out that $\frac{\partial u_n}{\partial \theta} = \frac{\partial u_n}{\partial z} \equiv 0$ on Γ_i. Hence $u_n = 0 \Rightarrow u_r = 0$ on Γ_i and hence $u_n \equiv 0 \rightarrow u_r \equiv 0$ on Γ_i and therefore $\mathbf{u}\cdot\nabla u_n = 0$. Further by a straightforward calculation it follows that $\mathbf{u}\cdot\nabla\mathbf{n}\cdot\mathbf{u} = \pm\mathbf{u}\cdot\nabla\mathbf{e}_\rho\cdot\mathbf{u} = \pm\frac{u_\theta^2(\rho_0, \theta, z)}{\rho_0}$. The results of the point c) are obtained in a completely analogous manner [13].

3. Energy relation for starshaped unbounded domains

Let $\mathcal{C}$ denote the class of starshaped convex domains having boundary constituted by unions of plane, cylindrical or spherical surfaces[2]. Let $\Omega_0 \subset \mathcal{C}$ and let Ω be the domain exterior to Ω_0. Our goal, in this section, is to obtain a suitable energy relation holding in Ω. For the sake of concreteness, we refer to the case that Ω_0 is a starshaped bounded domain constituted by the union of a cylinder and two half spheres of the same radius r_0 (Fig. 1).

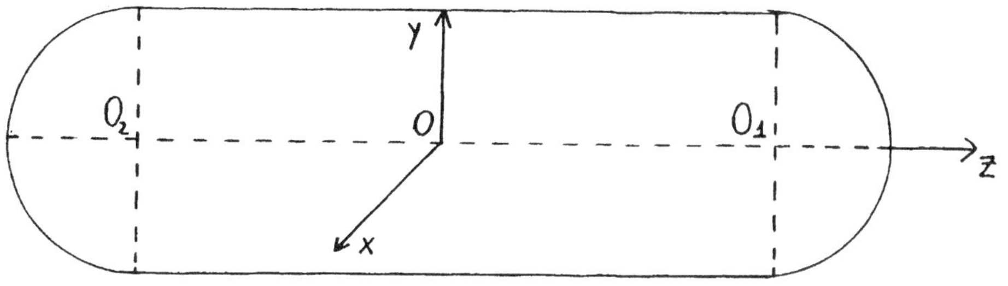

Figure 1

Denoting by O a point with respect to which Ω_0 is starshaped (Fig. 1), by C a "positive constant"[3] and setting $r = ||\mathbf{x}||$, the following theorem holds

[2]Of course $\mathcal{C}$ is not empty. In fact $\mathcal{C}$ contains the spheres, the cylinders, the cubes,...
[3]The numerical values of C is unessential and its values may be different in a single calculation.

Theorem 2 - *Let* $(\mathbf{u},\pi) \in \mathcal{P}$

$$\mathcal{P} = \{(\mathbf{u},\pi) : |\mathbf{u}| \leq C,\ |\pi|(1+r)^\gamma \leq C,\ \gamma > \frac{1}{2}\} \qquad in\ \Omega \times [0,T] \tag{3.1}$$

and

$$|\mathbf{v}(\mathbf{x},t)| \leq C\,. \tag{3.2}$$

Then, if

$$\exists t_0 \in [0,T) : \mathbf{u}(t_0) \in L^2(\Omega) \tag{3.3}$$

necessarily

$$\exists C(T) : \|\mathbf{u}(t)\|_2^2 + \int_0^T \|\nabla \mathbf{u}\|_2^2 d\tau \leq C \tag{3.4}$$

and the energy relation holds

$$\begin{aligned} \|\mathbf{u}(t)\|_2^2 &= \|\mathbf{u}(t_0)\|_2^2 + \\ &- 2\int_{t_0}^t \left[\int_\Omega \mathbf{u}\cdot\mathbf{D}\cdot\mathbf{u}\,d\Omega + \nu\|\nabla\mathbf{u}\|_2^2\right] d\tau + \\ &- \int_{t_0}^t \int_{\partial\Omega} u^2 \mathbf{v}\cdot\mathbf{n}\,d\sigma d\tau + \\ &- \frac{1}{r_0}\int_{t_0}^t \left[\sum_{i=1}^2 \int_{\Gamma_i} u^2 d\sigma + \int_{\Gamma_3} (\mathbf{u}\cdot\mathbf{e}_\theta)^2 d\sigma\right] d\tau \end{aligned} \tag{3.5}$$

where Γ_i $(i = 1,2)$ *are the spherical parts of the boundary while* Γ_3 *represents the cylindrical part. Further* $\mathbf{e}_\theta$ *is the unit vector in the* θ *direction of the cylindrical frame of reference* (O,ρ,θ,z).

Proof. Taking into account the points b)-c) of the Remark 1, the proof can be obtained by choosing $f = \varphi e^{-\alpha r}, \alpha = \text{const.} > 0$, where φ is given by (4.1) and following the procedures used in [12], [14][4].

Remark 2 - Let $\mathbf{v}_\infty$ be a constant vector. Taking into account $(2.5)_3$ and that

$$\begin{aligned} \nabla\cdot\mathbf{u} = 0 \Rightarrow \mathbf{u}\cdot\mathbf{D}\cdot\mathbf{u} &= \mathbf{u}\cdot\nabla\mathbf{v}\cdot\mathbf{u} = \mathbf{u}\cdot\nabla(\mathbf{v}-\mathbf{v}_\infty)\cdot\mathbf{u} = \\ &= \nabla\cdot\{[(\mathbf{v}-\mathbf{v}_\infty)\cdot\mathbf{u}]\mathbf{u}\} - \mathbf{u}\cdot\nabla\mathbf{u}\cdot(\mathbf{v}-\mathbf{v}_\infty)\,, \end{aligned} \tag{3.6}$$

---

[4]We precise that $\|\nabla\mathbf{u}\|_2^2 = \int_\Omega |\nabla\mathbf{u}|^2 d\Omega$, with $|\nabla\mathbf{u}|^2 = \nabla\mathbf{u} : \nabla\mathbf{u} = \sum_{i,j}^{1-3}\left(\frac{\partial u_i}{\partial x_j}\right)^2$. Further it is an easy matter to verify that $\frac{\partial f}{\partial n} \geq 0$ on Γ_i.

the relation (3.5) is equivalent to

$$\begin{aligned}
\|\mathbf{u}(t)\|_2^2 = \|\mathbf{u}(t_0)\|_2^2 + \\
+2\int_{t_0}^{t}\left[\int_{\Omega}\mathbf{u}\cdot\nabla\mathbf{u}\cdot(\mathbf{v}-\mathbf{v}_\infty)d\Omega-\nu\|\nabla\mathbf{u}\|_2^2\right]d\tau+ \\
-\int_{t_0}^{t}\int_{\partial\Omega}\mathbf{u}^2\mathbf{v}\cdot\mathbf{n}d\sigma d\tau+ \\
-\frac{1}{r_0}\int_{t_0}^{t}\left[\sum_{i=1}^{2}\int_{\Gamma_i}u^2d\sigma+\int_{\Gamma_3}(\mathbf{u}\cdot\mathbf{e}_\theta)^2d\sigma\right]d\tau
\end{aligned}\tag{3.7}$$

Remark 3 - From the relation (3.5), on letting $O_1O_2 \to 0$ (see Figure 1) it follows that Ω_0 becomes the sphere of radius r_0 centered at O and (3.5) becomes

$$\begin{aligned}
\|\mathbf{u}(t)\|_2^2 = \|\mathbf{u}(t_0)\|_2^2 + \\
-2\int_{t_0}^{t}\left[\int_{\Omega}\mathbf{u}\cdot\nabla\mathbf{v}\cdot\mathbf{u}d\Omega+\nu\|\nabla\mathbf{u}\|_2^2\right]d\tau+ \\
-\int_{t_0}^{t}\left[\int_{\partial\Omega}(\mathbf{v}\cdot\mathbf{n}+\frac{1}{r_0})u^2d\sigma\right]d\tau
\end{aligned}\tag{3.8}$$

which is essentially the energy relation obtained in [13] where the negative contribution linked to the presence of $\mathbf{u}\cdot\nabla\mathbf{n}\cdot\mathbf{u}$ on the boundary has been retained. Further, let $O_i = (0,0,z_i)$ and let $z_1 \to +\infty$, $z_2 \to -\infty$. Then Ω is the exterior of the cylindrical domain $\Omega_0 = \{\mathbf{x} = (\rho,\theta,z) \in I\!R^3,\, \rho \leq \rho_0,\, \theta \in [0,2\pi),\, z \in I\!R\}$ and (3.5) becomes

$$\begin{aligned}
\|\mathbf{u}(t)\|_2^2 = \|\mathbf{u}(t_0)\|_2^2 + \\
-2\int_{t_0}^{t}\left[\int_{\Omega}\mathbf{u}\cdot\mathbf{D}\cdot\mathbf{u}d\Omega+\nu\|\nabla\mathbf{u}\|_2^2\right]d\tau+ \\
-\int_{t_0}^{t}\left[\int_{\partial\Omega}(\mathbf{v}\cdot\mathbf{n}+\frac{1}{r_0})u^2d\sigma\right]d\tau+ \\
-\frac{1}{r_0}\int_{t_0}^{t}(\int_{\partial\Omega}u_\theta^2d\sigma)d\tau
\end{aligned}\tag{3.9}$$

Of course other two equivalent cases can be obtained from (3.5) on letting that only one of the points O_i tends to infinity.

4. Weighted inequalities for unbounded starshaped domains

In this section we obtain some weighted inequalities which will be used in obtaining conditions ensuring stability. We begin by recalling this immediate lemma

Lemma 1 - *Let a be a positive constant and let $\varphi : I\!R_+ \to I\!R_+$ denote the "cut-off" function*

$$\varphi(\xi) = \begin{cases} 1 & \xi \in [0,a] \\ 2\left(\dfrac{\xi}{a}\right)^3 - 9\left(\dfrac{\xi}{a}\right)^2 + 12\left(\dfrac{\xi}{a}\right) - 4 & \xi \in [a,2a] \\ 0 & \xi \in [2a,\infty)\,. \end{cases} \tag{4.1}$$

Then it follows that $\varphi \in C^1(I\!R_+)$ and $(\forall\, \xi \in I\!R_+)$

$$\begin{cases} \varphi(\xi) \geq 0, \quad \varphi'(\xi) \leq 0, \quad |\varphi'(\xi)| = O(\frac{1}{a}), \\ \lim\limits_{a\to\infty} \varphi(\xi) = 1, \quad \lim\limits_{a\to\infty} \varphi'(\xi) = 0\,. \end{cases} \tag{4.2}$$

Let $\Omega_0 \subset I\!R^3$ be a convex starshaped domain with respect to an inner point O. Precisely, let Ω_0 be starshaped with respect to O and such that $(P-O)\cdot \mathbf{n}_0 \geq 0$, $\forall\, P \in \partial\Omega_0$, where $\mathbf{n}_0$ is the outward unit normal to $\partial\Omega_0$. We begin by obtaining a weighted Poincaré inequality.

Theorem 3 - *Denoting by I the set of the vector valued functions $\mathbf{u} : \Omega \to I\!R^3$ such that*

$$\int_\Omega \left[\frac{u^2}{r^k} + \frac{1}{r^{k-2}}(\nabla\mathbf{u})^2\right] d\mathbf{x} + \int_{\partial\Omega} \frac{u^2}{r^{k-1}} d\sigma < \infty \quad k \in [2,3), r = |\mathbf{x}| \tag{4.3}$$

then $\forall\, \mathbf{u} \in I$, the following inequality holds

$$\int_\Omega \frac{u^2}{r^k} d\mathbf{x} \leq \left(\frac{2}{3-k}\right)^2 \int_\Omega \frac{|\nabla\mathbf{u}|^2}{r^{k-2}} d\mathbf{x} - \frac{2}{3-k}\int_{\partial\Omega} \frac{u^2}{r^k}(P-O)\cdot\mathbf{n}_0 d\sigma \tag{4.4}$$

Proof. Let (r,θ,ϕ) be a polar frame of reference with the origin at O and let φ denote the function (4.1) with $r = \xi$ and $\mathbf{e}_r = \nabla r$. From

$$\nabla\cdot\left[\frac{\varphi(r)u^2(P-O)}{r^k}\right] = \frac{\varphi' u^2}{r^{k-1}} + 2\varphi\frac{\mathbf{e}_r\cdot\nabla\mathbf{u}\cdot\mathbf{u}}{r^{k-1}} + (3-k)\frac{\varphi u^2}{r^k}$$

easily follows

$$(3-k-\varepsilon)\frac{\varphi(r)u^2}{r^k} \leq \nabla \cdot \left[\frac{\varphi(r)u^2(P-O)}{r^k}\right] + \\ -\varphi' r^{1-k}u^2 + \frac{1}{\varepsilon}\varphi r^{2-k}(\nabla \mathbf{u})^2 \tag{4.5}$$

where $\varepsilon(< 3-k)$ is a positive constant. Denoting by $S(a)$ the sphere centered at O and radius a and taking into account $(4.2)_3$, integrating it follows

$$(3-k-\varepsilon)\int_{\Omega\cap S(2a)} \varphi r^{-k}u^2 d\mathbf{x} \leq -\int_{\partial\Omega\cap S(2a)} r^{-k}u^2(P-O)\cdot \mathbf{n}_0 d\sigma + \\ +\frac{1}{\varepsilon}\int_{\Omega\cap S(2a)} \varphi r^{2-k}(\nabla \mathbf{u})^2 d\mathbf{x} + \\ +\frac{C}{a}\int_{\Omega\cap B(a)} r^{1-k}u^2 d\mathbf{x} \tag{4.6}$$

where $B(a) = S(2a) - S(a)$ and C is a positive constant associated to $(4.2)_3$. Taking into account (4.2)–(4.3) and letting $a \to \infty$, it turns out that

$$(3-k-\varepsilon)\int_{\Omega} r^{-k}u^2 d\mathbf{x} \leq \frac{1}{\varepsilon}\int_{\Omega} r^{2-k}(\nabla \mathbf{u})^2 d\mathbf{x} + \\ -\int_{\partial\Omega} r^{-k}u^2(P-O)\cdot \mathbf{n}_0 d\sigma \tag{4.7}$$

Choosing $\varepsilon = \dfrac{3-k}{2}$, (4.4) immediately follows.

Remark 4 - The value of the coefficient $\left(\dfrac{2}{3-k}\right)^2$ of $\| r^{(k/2)-1}|\nabla \mathbf{u}|\|_2^2$ appearing on the R.H.S. of (4.4) is of particular interest. In fact its reciprocal Rc represents a *critical value* for the Reynolds number Re of a basic motion, i.e. $Re < Rc$ ensures energy stability. Therefore (4.4), for $k \to 3$, has to be substituted by a different inequality holding also for $k = 3$.

Theorem 4 - *Let* $r_0 = \inf_{P\in\partial\Omega} \overline{OP} \geq e$ *and* $\{\sigma = const. \geq 0,\ n = const. \geq 1\}$. *Denoting by* I *the set of the vector valued function* $\mathbf{u} : \Omega \to \mathbb{R}^3$ *such that*

$$\begin{cases} \int_{\Omega}[(\sigma+\log r)^{n-1}r^{-k}u^2 + |\nabla \mathbf{u}|^2]d\Omega + \int_{\partial\Omega}(\sigma+\log r)^n r^{1-k}u^2 d\sigma < \infty \\ k \in (2,3]. \end{cases} \tag{4.8}$$

Then, $\forall \mathbf{u} \in I$, *the following inequality holds*

$$\int_{\Omega} \frac{(\sigma + \log r)^{n-1}}{r^k} u^2 d\Omega \le \frac{4e^{\sigma(k-2)}}{[n + (3-k)\sigma]^2} \left[\frac{n+1}{(k-2)e}\right]^{n+1} \int_{\Omega} (\nabla \mathbf{u})^2 d\Omega +$$
$$- \frac{2}{[n + (3-k)\sigma]} \int_{\partial\Omega} \frac{(\sigma + \log r)^n}{r^{k-1}} u^2 \mathbf{e}_r \cdot \mathbf{n}_0 d\sigma \tag{4.9}$$

Proof. From

$$\nabla \cdot \left[\frac{\varphi(r)(\sigma + \log r)^n u^2}{r^k}(P - O)\right] =$$
$$= \varphi[n + (3-k)(\sigma + \log r)] \frac{u^2}{r^k}(\sigma + \log r)^{n-1} +$$
$$+ \varphi'(r) \frac{(\sigma + \log r)^n}{r^{k-1}} u^2 + 2\varphi \frac{(\sigma + \log r)^n}{r^{k-1}} \mathbf{u} \cdot \frac{\partial \mathbf{u}}{\partial r}$$

easily follows that

$$\varphi(r)[n + (3-k)\sigma - \varepsilon] \frac{(\sigma + \log r)^{n-1}}{r^k} u^2 \le$$
$$\le \nabla \cdot \left[\frac{\varphi(r)(\sigma + \log r)^n u^2}{r^k}(P - O)\right] +$$
$$+ \frac{|\varphi'(r)|}{r^{k-1}}(\sigma + \log r)^n u^2 + \frac{1}{\varepsilon}\varphi \frac{(\sigma + \log r)^{n+1}}{r^{k-2}} |\nabla \mathbf{u}|^2 \tag{4.10}$$

where $\varepsilon (< n + (3-k)\sigma)$ is a positive constant.

But it is an easy matter to show that

$$\frac{(\sigma + \log r)^{n+1}}{r^{k-2}} \le \left[\frac{n+1}{(k-2)e}\right]^{n+1} e^{\sigma(k-2)} \qquad \forall \mathbf{x} \in \Omega. \tag{4.11}$$

Therefore integrating (4.10) on $S(a)$, letting $a \to \infty$ and choosing $\varepsilon = \frac{1}{2}[n + (3-k)\sigma]$, (4.9) immediately follows.

We mention as special case of the general inequality (4.9), the following one

Theorem 5 - *Let the assumptions of theorem 4 hold with* $k = 3$ *and* $n = 2$. *Then the following inequality holds*

$$\int_{\Omega} \frac{u^2}{r^3} d\Omega \le \left(\frac{3}{e}\right)^3 \int_{\Omega} (\nabla \mathbf{u})^2 d\Omega - \int_{\partial\Omega} \frac{\log^2 r}{r^2} u^2 \mathbf{e}_r \cdot \mathbf{n}_0 d\sigma \tag{4.12}$$

Proof. Choosing $k = 3$, $n = 2$ and $\sigma = 0$, from (4.9), (4.12) immediately follows.

Remark 5 - Let us remark that

i) Theorems 3-5 hold without any local assumption on the boundary. Further, standing the geometric properties of Ω_0, $\mathbf{e}_r \cdot \mathbf{n}_0 \geq 0$ on $\partial\Omega$ and hence the second integral appearing on the R.H.S. of (4.4), (4.9) and (4.12) is negative.

ii) the value of the coefficient $(3/e)^3$ of $||\nabla \mathbf{u}||_2^2$ appearing on the R.H.S. of (4.9) is less than 1.35.

5. Stability criteria

Coupling the energy relations (sect.3) to the weighted inequalities (sect.4), stability criteria for fluid motions occurring in the exterior of convex star-shaped domains, under stress free boundary like conditions can be obtained. As happens in the bounded domains [14] [3] and in the unbounded domains with homogeneous boundary data [12], the stability is linked to the supremum of the functional

$$\mathcal{F}(\mathbf{v}, \mathbf{u}, t) = \frac{-\int_\Omega \mathbf{u} \cdot \mathbf{D} \cdot \mathbf{u} d\Omega}{\int_\Omega \nabla \mathbf{u} : \nabla \mathbf{u} d\Omega}. \tag{5.1}$$

In fact from (4.5) it follows

$$\begin{aligned} ||\mathbf{u}(t)||_2^2 - ||\mathbf{u}(t_0)||_2^2 = 2 \int_{t_0}^t (\mathcal{F} - \nu) ||\nabla \mathbf{u}||_2^2 d\tau + \\ - \int_{t_0}^t \int_{\partial\Omega} u^2 \mathbf{v} \cdot \mathbf{n} d\sigma d\tau + \\ - \frac{1}{r_0} \int_{t_0}^t [\sum_{i=1}^2 \int_{\Gamma_i} u^2 d\sigma + \int_{\Gamma_3} (\mathbf{u} \cdot \mathbf{e}_\theta)^2 d\sigma] d\tau \end{aligned} \tag{5.2}$$

and one is led to consider the basic motions $(\mathbf{v}, p)$ for which $\mathcal{F}$ is bounded. Let the assumptions of theorems 2,3 hold and let $r_0 = \inf_{P \in \partial\Omega} |OP| > 0$, where O is a point with respect to which Ω_0 is starshaped. Exist classes of basic motions, physically meaningful, for which (5.1) is bounded in $\mathcal{P}$. This is the case, for instance, of the class $C_1(k)$

$$C_1(k) = \begin{cases} (\mathbf{v},p) : i)|\mathbf{v}| \le C, \ ii) m = \sup\limits_{t \ge t_0}\{\sup\limits_{\Omega} |\mathbf{D}| \left(\frac{r}{r_0}\right)^k\} < \infty, \\ k \in [2,3]; \ iii) \left(\frac{4mr_0^k}{3-k}\mathbf{e}_r - \mathbf{v}\right) \cdot \mathbf{n} \le 0 \text{ a.e. on } \partial\Omega, \ \forall t \ge t_0 \end{cases}$$

Let $(\mathbf{v},p) \in C_1(k)$. Then it follows

$$(\mathcal{F} - \nu)\|\nabla \mathbf{u}\|_2^2 \le mr_0^k \int_\Omega \frac{u^2}{r^2} d\Omega - \nu \int_\Omega |\nabla \mathbf{u}|^2 d\Omega \tag{5.3}$$

hence, by (4.4)

$$(\mathcal{F} - \nu)\|\nabla \mathbf{u}\|_2^2 \le [mr_0^2\left(\frac{2}{3-k}\right)^2 - \nu]\|\nabla \mathbf{u}\|_2^2 - \frac{2mr_0^k}{3-k}\int_{\partial\Omega} u^2 \nabla \mathbf{x} \cdot \mathbf{n}_0 d\sigma . \tag{5.4}$$

Therefore from (3.5) it follows

$$\begin{aligned} \|\mathbf{u}(t)\|_2^2 - \|\mathbf{u}(t_0)\|_2^2 \le 2[mr_0^2\left(\frac{2}{3-k}\right)^2 - \nu] \int_{t_0}^t \|\nabla \mathbf{u}\|_2^2 d\tau + \\ + \int_{t_0}^t \int_{\partial\Omega} \left(\frac{4mr_0^k}{3-k}\mathbf{e}_r - \mathbf{v}\right) \cdot \mathbf{n} u^2 d\sigma d\tau + \\ - \frac{1}{r_0}\int_{t_0}^t [\sum_{i=1}^2 \int_{\Gamma_i} u^2 d\sigma + \int_{\Gamma_3} (\mathbf{u}\cdot\mathbf{e}_\theta)^2 d\sigma] d\tau . \end{aligned} \tag{5.5}$$

Introducing the Reynolds number

$$Re = \frac{mr_0^2}{\nu} \tag{5.6}$$

it follows that *the condition*

$$Re < \left(\frac{3-k}{2}\right)^2 \tag{5.7}$$

implies stability.

Remark 6 - The case Ω_0 = sphere centered at O is of particular interest. In fact in this case

$$\Gamma_3 = \emptyset, \quad \partial\Omega = \Gamma_1 \cup \Gamma_2, \quad r = r_0, \quad \mathbf{n} = -\mathbf{e}_r$$

and conditions iii) concerning the basic motions becomes

$$\frac{4mr_0^k}{3-k} + \frac{1}{r_0} + v_n \geq 0 \quad \text{a.e. on } \partial\Omega, \forall t \geq t_0$$

Remark 7 - A stability condition analogous to (5.7) can be obtained by using the energy relation (3.7).

Remark 8 - The class of the basic motions is not empty. In fact contains at least the steady motions

$$(\mathbf{v}, p) \equiv \left(\frac{Q}{4\pi r^2} \nabla \mathbf{x}, \; -\frac{Q^2}{32\pi^2 r^4} + U + \text{cost.} \right) \tag{5.8}$$

where the body force acting on the fluid is $\mathbf{F} = \nabla U$ [15].

Exploiting the condition (5.7) to the basic motion (5.8), the stability results of [13] can be immediately generalized to the exterior of any convex starshaped domain Ω_0 for which inequality (4.4) holds. But, taking into account the inequality (4.12), this stability results can be improved. For the sake of simplicity, we consider here the basic motion (5.8) in the case $Q < 0$ (i.e. O is a sink) and set

$$V = 4\pi \max_{\Omega}[|\mathbf{v}| \left(\frac{r}{r_0}\right)^2], \quad Re = \frac{V r_0}{\nu} = \frac{|Q|}{\nu r_0}.$$

It turns out that

$$V = \frac{Q}{r_0^2}, \quad |\mathbf{v} - \mathbf{v}_\infty| \leq \frac{V}{4\pi}\left(\frac{r_0}{r}\right)^2 = \frac{V r_0^2}{4\pi r^2} = \frac{\nu r_0}{4\pi r^2} Re$$

$$2\mathbf{u} \cdot \nabla \mathbf{u} \cdot |\mathbf{v} - \mathbf{v}_\infty| \leq \frac{u^2}{\nu} |\mathbf{v} - \mathbf{v}_\infty|^2 + \nu |\nabla \mathbf{u}|^2 \leq \frac{\nu r_0^2 Re^2}{(4\pi)^2} \frac{u^2}{r^4} + \nu (\Delta \mathbf{u})^2.$$

Taking into account that (4.12) implies

$$r_0^2 \int_\Omega \frac{u^2}{r^4} d\Omega \leq \left(\frac{3}{e}\right)^3 \int_\Omega (\nabla \mathbf{u})^2 d\Omega,$$

where now the L.H.S. and the R.H.S. have the same dimensions, from (3.7) then it follows

$$\|\mathbf{u}(t)\|_2^2 \leq \|\mathbf{u}(t_0)\|_2^2 + \nu[\left(\frac{Re}{4\pi}\right)^2\left(\frac{3}{e}\right)^3 - 1]\int_{t_0}^{t} \|\nabla\mathbf{u}\|_2^2 d\tau\,.$$

Therefore $Re < Rc$ with $Rc = 4(e/3)^{3/2}\pi > 3.4\pi$ ensures the stability of the basic flow (5.8) in the case $Q < 0$.

Remark 9 - As is well known the best energy stability conditions are linked to the variational problem associated to the existence of the supremum of the functional (5.1)

Acknowledgements - The supports of the Italian Ministry for University and Scientific Research (M.U.R.S.T. 40 % and 60 % contracts) and the project "Applicazioni della matematica per la tecnologia e la societá" of the italian C.N.R., are acknowledged.

References

1. Solonnikov V.A. and Scadilov V.E.: On the boundary value problem for stationary system of Navier-Stokes equations, *Proc.Steklov Ins. Math.* **125** (1973), 186-199.
2. Solonnikov V.A.: Solvability of a problem on the motion of a viscous incompressible fluid bounded by a free surface, *Math. USSR Izv.* **11** (1977), 1323-1358.
3. Solonnikov V.A.: Solvability of a three-dimensional boundary value problem with a free surface for the stationary Navier-Stokes system, *Banach Center Publication PWN, Warsaw.* **10** (1983), 361-403.
4. Joseph D.D.: *Stability of Fluid Motion*, vols. I- II, Springer, Berlin, 1976.
5. Rionero S.: Sulla stabilitá magnetoidrodinamica con vari tipi di condizioni al contorno, *Ric. Mat.* **17** (1968), 64-78.
6. Fabes E.B., Lewis J.E. and Riviere M.B.: Boundary value problems for the Navier-Stokes equations, *Ann. J. Math.* **99** (1977), 628-668.
7. Rivkind V. YA and Fridmann N.M.: The Navier-Stokes equations with discontinuous coefficients, *J. Sov. Math.* **58** (1977), 456-529.
8. Mulone G. and Salemi F.: On the existence of hydrodynamic motion in a domain with free boundary type conditions, *Meccanica* **18** (1983), 136-144.

9. Mulone G. and Salemi F.: On the hydrodynamic motion in a domain with mixed boundary conditions: existence, uniqueness, stability and linearization principle, *Ann. Mat. Pura Appl.* **139** (1985), 147- 174.
10. Rionero S. and Mulone G.: Existence and uniqueness theorems for a steady thermo-diffusive mixture in a mixed problem, *Nonlinear analysis, Theory Methods and Applications* **12** (1988), n.5, 473-494.
11. Rionero S. and Mulone G.: On the nonlinear stability of a thermodiffusive fluid mixture in a mixed problem, *J. Math. Anal. and Appl.* (1) **124** (1987), 165- 188.
12. Galdi G.P. and Rionero S.: *Weighted energy methods in fluidynamics and elasticity*, Lecture Notes in Mathematics **1134**, Springer-Verlag.
13. Capone F. and De Angelis M.: On the energy stability of fluid motions in exterior of a spere under free boundary like conditions, *Rend. Acc. Sc. Fis. Mat.* Serie IV **Vol. LX** (1993).
14. Serrin J.: On the stability of fluid motions, *Arch. Rat. Mech. Anal.* **3** (1959), 1-13.
15. Berker R.: Integration des equations du mouvement d'un fluide visqueux incompressible, *Handbuch der physik*, **Bd VIII/2** (1963), 1-384.

VARIATIONAL LIMIT OF COMPRESSIBLE TO INCOMPRESSIBLE FLUID

LISA SANTOS
Departamento de Matemática
Universidade do Minho
Campus de Gualtar
4710 Braga
Portugal

Abstract

We consider the flow with (or without) wake of a stationnary, irrotational, compressible fluid, with non-constant density, in a channel (or in the whole plane), with given velocity at infinity and at the wake.

Considering a sequence of densities ρ_n and letting $\rho_n \longrightarrow 1$, when $n \to \infty$, we are going to prove that the variational solution u_n of the compressible problem converges (in a sense that will be defined) to the variational solution of the incompressible problem.

Although the linear variational inequalities corresponding to density $\rho_n \neq 1$ have non-constant coefficients, it is still possible to show that the free boundaries are graphs of Lipschitz functions. Calling l_n and l the graphs of the free boundaries of problems corresponding to densities ρ_n and 1 respectively, we prove that $l_n \longrightarrow l$ in a convenient space, when $n \to \infty$.

We also obtain a result of continuous dependence on the data, of the solutions.

1 The physical problem and the variational inequality

Consider a flow with (or without) wake of a stationnary, irrotational, compressible fluid, with non-constant density ρ, in a channel (or in the whole plane), with given velocity q_∞ at infinity and $q_s < q_\infty$ on the wake, around a convex profile with height $2H$, symmetric with respect to the OX axe.

S. N. Antontsev et al. (eds.), Energy Methods in Continuum Mechanics, 126–144.

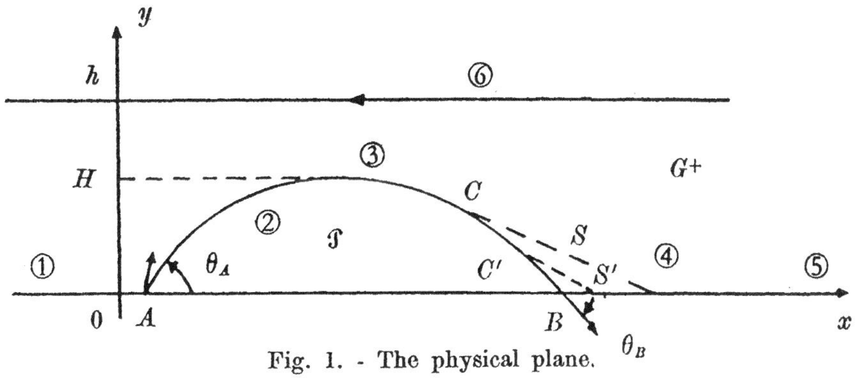

Fig. 1. - The physical plane.

The flow is supposed to be uniform at infinity and the velocity at infinity to be parallel to the OX axe. The given density of the fluid is a function of q, where q is the scalar velocity.

More precisely, we impose that

$$\rho = g(q) \quad g \text{ is a decreasing positive } C^1 \text{ function.} \tag{1}$$

These kind of problems were treated before. In 1973, Brézis and Stampacchia studied the problem of an incompressible fluid in the plane ([3]) and Brézis and Duvaut ([2]) the problem of an incompressible fluid with wake, in the plane; in1975 Brézis ([1]) considered the problem of a compressible fluid in the plane; in 1978 Tomarelli treated the case of an incompressible fluid in a channel; in 1978 Shimborsky ([8], [9]) studied incompressible fluids in a the symetric plane channel with convex boundaries or in Venturi channels; in 1982 Díaz ([4]) applied the technique of local supersolution to the study of stationnary nonlinear problems and at the same year Hummel used the hodograph method applied to the study of problems involving convex profiles. In 1988 the author considered the asymptotic behavior of the solutions and free boundaries of problems of an incompressible fluid with wake in a channel, when the velocity of the wake tends to zero or the height of the channel tends to infinity ([6]) and in 1993 the same questions where solved, now for compressible fluids, with the aditional asymptotic convergence of the solutions of problems of compressible fluids to the solution of the problem of the incompressible one, when the density of the fluids tends to 1 ([7]). The present work gives a syntheses of this last question, with an improvement on the assumptions on the densities and with an original result concerning the continuous dependence on the data (namely, on the geometry of the profile, on the velocity at infinity and on the wake) of the solutions.

We are going to restrict ourselves to the case

- Compressible fluid,
- in a channel with semi-height h,
- with wake, with constant velocity q_s at the wake and q_∞ at infinity.

We will rapidly obtain the physical formulation of the problem and after using a change of variables, the hodograph transformation, and a change of variables of Baiocchi type, this problem becomes a variational inequality problem; letting $(\rho_n)_{n\in I\!N}$ be a sequence of densities converging to 1, we will prove that the solutions $(u_n)_{n\in I\!N}$ of the problem (in the hodograph plane) will converge to the solution of the incompressible problem, as well as the free boundaries. In last section we obtain the continuous dependence result.

Let $\mathcal{P}$ denote the profile and $\mathcal{G}$ the exterior of the profile in the channel. Define

$$\begin{aligned} \mathcal{G}^+ &= \mathcal{G}\cap\{(x,y)\in I\!R^+ : y\geq 0\} \\ \mathcal{P}^+ &= \mathcal{P}\cap\{(x,y)\in I\!R^2 : y\geq 0\} \end{aligned}$$

and, if $\vec{q}=(q_1,q_2)$ denotes the velocity of the flow, let ψ be stream function, defined by

$$\psi_x=-\rho q_2 \qquad \psi_y=-\rho q_1,$$

Our problem is to find ψ such that

$$\left\{\begin{array}{l} \psi\in C^2(G^+)\cap C^1(\overline{G}^+), \\ \left(1-\dfrac{q_1^2}{a^2(q)}\right)\psi_{xx}+\left(1-\dfrac{q_1^2}{a^2(q)}\right)\psi_{yy}-2\dfrac{q_1q_2}{a^2(q)}\psi_{xy}=0 \qquad \text{in } G, \\ |\nabla\psi|=q_sg(q_s) \qquad \text{on } S, \\ \psi(x,h)=hq_\infty g(q_\infty) \qquad \forall x\in I\!R, \\ \psi-0 \qquad \text{on } \partial\mathcal{P}^+\cup S \text{ or if } y=0, \\ \lim_{|x|\to+\infty}\psi(x,y)=yq_\infty g(q_\infty) \qquad \text{uniformly in } y, \end{array}\right. \tag{2}$$

where $a^2(q) = -q\dfrac{g(q)}{g'(q)}$. We suppose that the equation $a(q)=q$ has a positive solution and denote by q_c (velocity of the sound) the least positive solution of $a(q)=q$. The equation of problem (2)

- is elliptic in the subsonic domain $\{q<q_c\}$;
- hiperbolic in the supersonic domain $\{q>q_c\}$.

We restritct ourselves to the elliptic case. Considering x and y as functions of q and θ ($q_1=q\cos\theta, q_2=q\sin\theta, q=\|(q_1,q_2)\|$) and defining

$$\sigma=\int_q^{q_c}\frac{g(\tau)}{\tau}d\tau=\sigma(q) \tag{3}$$

$$\sigma_\infty=\int_{q_\infty}^{q_c}\frac{g(\tau)}{\tau}d\tau, \qquad s=\int_{q_s}^{q_c}\frac{g(\tau)}{\tau}d\tau, \qquad \sigma_h=\int_{q_h}^{q_c}\frac{g(\tau)}{\tau}d\tau, \tag{4}$$

where $q_h = \max_{x \in I\!R} |\vec{q}(x,h)|$ and letting

$$R(\theta) = -[(X'(\theta))^2 + (Y'(\theta))^2]^{\frac{1}{2}} \tag{5}$$

where $(X(\theta), Y(\theta))$ is the unique point P of $\partial\mathcal{P}^+$ where the tangent at P to $\partial\mathcal{P}^+$ makes an angle θ ($\theta < \pi/2$) with the axe OX.

Assuming that

$$X, Y \in C^{1,\alpha}(\theta_{B,}\theta_A), \qquad 0 < \alpha < 1, \tag{6}$$

it can be seen that problem (2) is equivalent to problem (7) (in the coordinates (θ,σ)) : to find Ψ defined in Ω_s and $\mathcal{D} \subseteq \Omega_s$ such that

$$\begin{cases} k(\sigma)\Psi_{\theta\theta} + \Psi_{\sigma\sigma} = 0 & \text{in } \mathcal{D}, \\ \begin{cases} \Psi = 0 \\ \Psi_\sigma = -\frac{R(\theta)q(\sigma)}{1+k(\sigma)\ell'(\theta)} \end{cases} & \text{on } \mathcal{L}, \\ \Psi = 0 & \text{on } \Gamma_s \cup \Sigma_\infty, \\ \Psi = hq_\infty g(q_\infty) & \text{on } \Sigma_h, \\ \Psi \to 0 & \text{when } \sigma \to \infty, \quad \theta > 0, \end{cases} \tag{7}$$

where

$$\Omega_s =]\theta_B, 0] \times]0, s[\cup]0, \theta_A[\times I\!R^+ \tag{8}$$

and

$$k(\sigma) = \frac{1}{g^2(q(\sigma))}\left(1 - \frac{q^2(\sigma)}{a^2(q(\sigma))}\right), \tag{9}$$

$$\begin{aligned} \mathcal{D} = \{(\theta,\sigma) : \theta_s < \theta < \theta,\ \ell(\theta) < \sigma < s\} \cup \{(\theta,\sigma) : 0 < \theta < \theta_A,\ \ell(\theta) < \sigma\} \\ \cup\{0,\sigma) : \ell(0) < \sigma < \sigma_h\}, \end{aligned} \tag{10}$$

$$\Gamma_s = \{(\theta, s) : \theta_s < \theta < 0\}, \tag{11}$$

$$\Sigma_h = \{(0,\sigma) : \sigma_h < \sigma < \sigma_\infty\}, \qquad \Sigma_\infty = \{(0,\sigma) : \sigma_\infty < \sigma < +\infty\}. \tag{12}$$

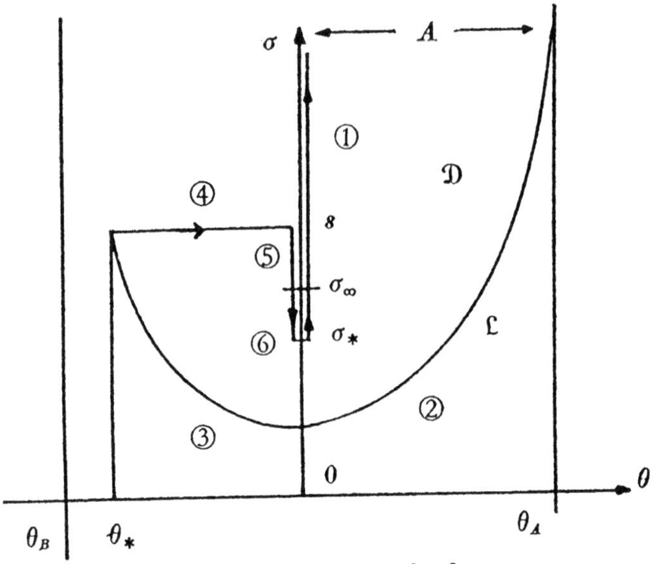

Fig. 2. - The hodograph plane.

Introduce now a change of variable of Baiocchi type

$$u(\theta,\sigma) = \begin{cases} \int_{\ell(\theta)}^{\sigma} \frac{k(\tau)}{q(\tau)} \Psi(\theta,\tau) d\tau & \text{if } (\theta,\sigma) \in \mathcal{D} \\ 0 & \text{if } (\theta,\sigma) \in \Omega_s \setminus \mathcal{D}. \end{cases} \tag{13}$$

Let

$$\Gamma =]\theta_B, 0] \times \{s\}$$

$$V_s = \{v : \; qv \in L^2(\Omega_s), \; qv_\theta \in L^2(\Omega_s), \; \tfrac{q}{\sqrt{k}} v_\sigma \in L^2(\Omega_s), \\ v|_{\partial\Omega_s \setminus \Gamma} \equiv 0\}$$

$$\|v\|_{V_s}^2 = \int_{\Omega_s} q^2 \left(\frac{1}{k} v_\sigma^2 + v_\theta^2 + v^2 \right) d\theta d\sigma \tag{14}$$

Define on V_s the bilinear symmetric form

$$a(u,v) = \int_{\Omega s} q^2 \left(\frac{1}{k} u_\sigma v_\sigma + u_\theta + v_\theta - uv \right) d\theta d\sigma \tag{15}$$

and the convex set

$$I\!K = \{v \in V_{\cdot} \; v \geq 0, \; v(0,\sigma) = H \text{ if } \sigma \geq \sigma_\infty, \; v(0,\sigma) \geq \eta(\sigma) \text{ if } \sigma < \sigma_\infty\}, \tag{16}$$

where

$$\eta(\sigma) = H - hq_\infty g(q_\infty) \int_\sigma^{\sigma_\infty} \frac{k(\tau)}{q(\tau)} d\tau. \tag{17}$$

Then u is the unique solution of the variational inequality

$$\left\{ \begin{array}{l} u \in \mathbb{K} \\ a(u, v-u) \geq \int_{\Omega_s} q^2(\sigma) R(\sigma)(v-u) d\theta d\sigma, \quad \forall v \in \mathbb{K}. \end{array} \right. \tag{18}$$

We can think about the variational inequality itself, being $q(\sigma)$ and $k(\sigma)$ defined by

$$\begin{array}{ll} \frac{q'(\sigma)}{q(\sigma)} = -\frac{1}{g(q(\sigma))}, & q(0) = 1 \\ k(\sigma) = \frac{1}{g^2(q(\sigma))}\left[1 + \frac{q(\sigma) g'(q(\sigma))}{g(q(\sigma))}\right] & \end{array} \tag{19}$$

assuming that $g \in W^{3,\infty}(0, q_c)$, in order to garantee the boundedness of the first and second derivatives of q and k. Notice that,

$$0 \leq q(\sigma) < q_c, \qquad k(\sigma) \geq 0,$$

- σ_∞ is given, $q_\infty = q(\sigma_\infty)$; $\quad s\ (> \sigma_\infty)$ is given, $q_s = q(s)$;
- $\mathcal{D} = \{(\theta, \sigma) \in \Omega_s : u(\theta, \sigma) > 0\}$ ($\mathcal{D}$ is open);
- $\sigma_* = \inf\{\sigma : u(\theta, \sigma) = H - h q_\infty g(q_\infty) \int_\sigma^{\sigma_\infty} \frac{k(\tau)}{q(\tau)} d_\tau\}$, $\theta_* = \inf\{\theta : u(\theta, s) > 0\}$;
- $\Sigma_\infty = \{0\} \times [\sigma_\infty, +\infty[$, $\quad \Sigma_* = \{0\} \times]\sigma_*, \sigma_\infty[$, $\quad \Gamma_* =]\theta_*, 0[\times \{s\}$.

We begin with a regularity result.

Theorem 1.1
1) *For all* $\alpha, 0 < \alpha < 1$ *and some* $\beta, 0 < \beta < 1$,

$$u \in C^{2,\alpha}(\mathcal{D} \setminus \Sigma_\infty \cup \Sigma_* \cup \Gamma_*),$$

$$u \in C^{1,\alpha}(\Omega_A \cup \Sigma_\infty), \quad u \in C^{1,\alpha}(\Omega_B \cup \Sigma_\infty \cup \Gamma_*),$$

$$u \in C^{1,\beta}(\Omega_A \cup \Sigma_*), \quad u \in C^{1,\beta}(\Omega_B \cup \Sigma_*).$$

2) *The free boundary* $\mathcal{L} = \Omega_s \cap \partial\{(\theta, \sigma) : u(\theta, \sigma) = 0\}$ *is a graph* $\sigma = \ell(\theta)$, *uniformly Lipschiz on any compact subset of* $]\theta_*, \theta_A[$. *The Lipschitz constant depends on* $\|u\|_{C^1(K)}$ (*being* K *any compact subset of* Ω_s), $\max_{\theta \in [\theta_B, \theta_A]} |R(\theta)|$, $\max_{\theta \in [\theta_B, \theta_A]} |R'(\theta)|$, $\|q\|_{W^{2,\infty}}$, $\|K\|_{W^{2,\infty}}$.
3) $\ell \in C^{1,\alpha}_{loc}$ *and* $u \in C^{1,\alpha}(D \cup \mathcal{L})$.

Proof: The proof is ommited. It can be found in detail in [7]. ∎

2 The incompressible case as a limit

In this section we are going to prove that the solution of the variational inequality of the problem of an incompressible fluid is the limit of the solutions of the variational inequalies of the problems of compressible fluids with densities ρ_n, when $\rho_n \to 1$ when $n \to +\infty$.

Let $(g_n)_n$ be a sequence of densities (of the fluid) such that

$$\begin{cases} g_n \in W^{3,\infty}(\mathbb{R}_0^+) \\ g_n \longrightarrow 1, \; g_n' \longrightarrow 0 \quad \text{uniformly on the compact subsets of } \mathbb{R}_0^+, \\ \qquad\qquad\qquad\qquad \text{when } n \to +\infty. \end{cases} \tag{20}$$

Remark 2.1 *In* [7] we imposed that

$$\exists m > 0 \; \forall n \in \mathbb{N} \; \forall x \in \mathbb{R}_0^+ \quad g_n(x) \geq m.$$

This condition is ommited here, since it is not necessary. ■

For a given g_n, let us define

$$\xi_n = \int_1^{q_c^n} \frac{g_n(\tau)}{\tau} d\tau \tag{21}$$

where q_c^n is the velocity of the sound, i.e., the least positive solution of $\frac{g_n(x)}{g_n'(x)} = -x$. For n suficiently large, ξ_n is positive, since it can be proved (see [7]) that $q_c^n \longrightarrow +\infty$ when $n \to +\infty$.

Define

$$s_n = \int_{q_s}^1 \frac{g_n(\tau)}{\tau} d\tau, \qquad \sigma_n = \int_{q_\infty}^1 \frac{g_n(\tau)}{\tau} d\tau \tag{22}$$

Let q_n and k_n be defined, as before, by the equations

$$\frac{q_n'(\sigma)}{q_n(\sigma)} = -\frac{1}{g_n(q_n(\sigma))}, \qquad q_n(\xi_n) = 1$$

$$k_n(\sigma) = \frac{1}{g_n^2(q_n(\sigma))}\left(1 + \frac{q_n(\sigma) g_n'(q_n(\sigma))}{g_n(q_n(\sigma))}\right).$$

The variable σ (which depends on n) is related with q_n (taken as a variable) by the following relation: $\sigma(q_n) = \displaystyle\int_{q_n}^{q_c^n} \frac{g_n(\tau)}{\tau} d\tau$. Redefining the problem with a different initial condition for q_n corresponds to make a translation on the variable

σ, i.e. , $\sigma \rightsquigarrow \sigma + \xi_n$ and, in the new variables, $\sigma(q_n) = \int_{q_n}^{1} \frac{g_n(\tau)}{\tau} d\tau$, q_n is defined in $]-\xi_n, +\infty[$ and $q_n(0) = 1$.

Let

$$\begin{cases} \Gamma_{s_n} =]\theta_B, 0[\times\{s_n\} \\ \Omega_{s_n} =]\theta_B, 0]\times]-\xi_n, s_n[\cup]0, \theta_A[\times]-\xi_n, +\infty[. \end{cases} \tag{23}$$

Everything that was done for $n \in \mathbb{N}$ is now also extensible for $n = +\infty$, letting $g_\infty \equiv 1$. Notice that $q_\infty(\sigma) = e^{-\sigma}$ and $k_\infty(\sigma) \equiv 1$, for any $\sigma \in \mathbb{R}$. Let $n \in \mathbb{N} \cup \{+\infty\}$,

$$V_n = \{v : q_n v \in L^2(\Omega_{s_n}),\ q_n v_\theta \in L^2(\Omega_{s_n}),\ \frac{q_n}{\sqrt{k_n}} v_\sigma \in L^2(\Omega_{s_n}), \tag{24}$$

$$v_{|\partial\Omega_{s_n}\setminus\Gamma_{s_n}} \equiv 0\} \tag{25}$$

and

$$\|v\|_n^2 = \int_{\Omega_{s_n}} q_n^2 \left(\frac{1}{k_n} v_\sigma^2 + v_\theta^2 + v^2\right) d\theta d\sigma. \tag{26}$$

Recall that

$$\eta_n(\sigma) = H - Hhq_\infty g_n(q_\infty) \int_\sigma^{\sigma\infty} \frac{k_n(\tau)}{q_n(\tau)} d\tau \tag{27}$$

and the convex sets

$$\mathbb{K}_n = \{v \in V_n : v \geq 0,\ v(0,\sigma) = H \text{ if } \sigma \geq \sigma_n, \tag{28}$$

$$v(0,\sigma) \geq \eta_n(\sigma) \text{ if } \sigma < \sigma_n\}. \tag{29}$$

Consider the bilinear form

$$a_n(u,v) = \int_{\Omega_{s_n}} q_n^2 \left(\frac{1}{k_n} u_\sigma v_\sigma + u_\theta v_\theta - uv\right) d\theta d\sigma \tag{30}$$

The solution of our problem in the compressible case with density function g_n (this includes the incompressible case with $g_\infty \equiv 1$) is the unique solution of the variational inequality

$$\begin{cases} v_n \in \mathbb{K}_n, \qquad n \in \mathbb{N} \cup \{+\infty\} \\ a_n(v_n, v - v_n) \geq \int_{\Omega_{s_n}} R(\theta) q_n^2(\sigma)(v - v_n) d\theta d\sigma, \qquad \forall v \in \mathbb{K}_n. \end{cases} \tag{31}$$

Theorem 2.2 ([1])
Consider problem (18) *with density function* g. *Let* $\sigma_* = inf\{\sigma : \eta(\sigma) > 0\}$ *and* $q(\sigma_*) = q_*$. *Let* $\alpha = \min_{\theta\in[\theta_B,\theta_A]} |R(\theta)| > 0$ *and* q_μ *be the least positive solution of the equation*

$$\frac{q_\mu}{q_*}\left[-1 + \frac{1}{g(q_*)}\int_{q_*}^{q_\mu} \frac{g(\tau)}{\tau} d\tau\right] = \frac{H}{\alpha} - 1. \tag{32}$$

If $q_\mu \leq q_c$, *then the maximum velocity of the fluid is less or equal to* q_μ.

Proof: See [1]. ∎

Remark 2.3 *The above theorem is valid whenever* $q_\mu \leq q_c$. *When* $g_n \longrightarrow 1$ *as* $n \to +\infty$ *(uniformly on the compact subsets) then* $q_c^n \longrightarrow +\infty$. *On the other hand,* $q_\mu^n \longrightarrow q_\mu$, *where* q_μ *is the solution of equation* (32) *with density function* $g \equiv 1$. *Then, at least for* n *big, the preceding theorem is true in our case.*

Then

$$\exists n_0 \in \mathbb{N}\ \exists \mu \in \mathbb{R} : \quad \begin{cases} \forall n \in \mathbb{N} \quad \mu > -\xi_n, \\ \forall n \in \mathbb{N}\ \forall \sigma \leq \mu \quad v_n(\theta,\sigma) = 0. \end{cases}$$

Consider then the open sets

$$\tilde{\Omega}_{sn} =]\theta_B, 0] \times]\mu, s_n[\cup]0, \theta_A[\times]\mu, +\infty[\tag{33}$$

Let

$$\tilde{v} = v_{|\tilde{\Omega}_{s_n}}, \qquad \tilde{V}_n, \quad \tilde{\mathbb{K}}_n, \quad \tilde{a}_n$$

be defined in the natural way, considering the intervenient functions with domain of definition $\tilde{\Omega}_{s_n}$.

Then, noticing that $R < 0$, we easily verify that $\tilde{v}_n$ is the unique solution of the variational inequality

$$\begin{cases} \tilde{v}_n \in \tilde{\mathbb{K}}_n, \\ a_n(\tilde{v}_n, v - \tilde{v}_n) \geq \int_{\tilde{\Omega}_{sn}} q_n^2(\sigma) R(\theta)(v - \tilde{v}_n) d\theta d\sigma, \qquad \forall \tilde{v} \in \tilde{\mathbb{K}}_n. \end{cases} \tag{34}$$

Remark 2.4 *Since there exists no risk of confusion, we will ommit, from now on, the* $\tilde{}$.

Lemma 2.5

1) $q_n(\sigma) = e^{-\int_0^\sigma \frac{d\tau}{g_n(q_n(\tau))}}$;

2) $q_n \in C^3([\mu, +\infty[)$ *and* $k_n \in C^1([\mu, +\infty[)$;

3) $q_n \longrightarrow e^{-\sigma}, q_n' \longrightarrow_n -e^{-\sigma}$, *uniformly on the compact subsets of* $[\mu, +\infty[$ *and* $q_n \longrightarrow e^{-\sigma}$ *in* $H^1(\mu, +\infty)$, *when* $n \to +\infty$;

4) $k_n \longrightarrow 1$, $k_n' \longrightarrow 0$, *uniformly on the compact subsets of* $[\mu, +\infty[$ *and* $k_n \longrightarrow 1$ *in* $H^1(\mu, +\infty)$, *when* $n \to +\infty$.

Proof: See [7]. ∎

The problem for an incompressible fluid is defined as follows (see [3]): Let

$$\Omega_s' =]\theta_B, 0] \times]\beta, s[\cup]0, \theta_A[\times]\beta, +\infty[,$$

where $s = -\log q_s$ and β is any lower bound of $\tau_h + \sigma_\infty + \log(1 - H/h)$, where τ_h is the unique negative solution of the equation $(1 + \frac{H}{\rho})(1 - \frac{H}{h}) = e^\tau(\tau + 1 + \log(1 - \frac{H}{h}))$, being $\rho = \max_{\theta \in [\theta_B, \theta_A]} R(\theta) < 0$. Let

$$\Gamma_s =]\theta_B, 0[\times \{s\} \qquad V' = \{v \in H^1(\Omega_s') : v_{|\partial\Omega_s' \setminus \Gamma_s} \equiv 0\}$$

$$\mathbb{K}' = \{v \in V' : v \geq 0,\ v(0,\sigma) = He^{-\sigma} \text{ if } \sigma > \sigma_\infty,\ v(0,\sigma) \geq \Phi(\sigma) \text{ if } \sigma < \sigma_\infty\}$$

where $\Phi(\sigma) = hq_\infty + (H - h)e^{-\sigma}$ and define the bilinear form

$$b(u,v) = \int_{\Omega_s'} \nabla u . \nabla v d\theta d\sigma + \int_{\Gamma_s} uv d\theta. \tag{35}$$

Then, the solution u of the problem for an incompressible fluid, formulated in the hodograph plane, using variational inequalities, is the unique solution of

$$\begin{cases} u \in \mathbb{K}', \\ b(u, v-u) \geq \int_{\Omega_s'} R(\theta)e^{-\sigma}(v-u)d\theta d\sigma, \qquad \forall v \in \mathbb{K}'. \end{cases} \tag{36}$$

Remark 2.6 *Consider the problem* $n = +\infty$ *and let* v_∞ *be the solution of that problem.*

i) We can consider $\beta = \mu$ *extending* u *or* v_∞ *by zero, whether we have* $\beta \geq \mu$ *or* $\beta < \mu$*; since to* $\mathbb{K}$ *and* $\mathbb{K}'$ *only belong non-negative functions and* $R < 0$*, these extensions* u *or* v_∞ *are the solutions of the variational inequality in the extended domain.*

ii) Given v, $w \in V'$ *notice that* $e^\sigma v$, $e^\sigma w \in V_\infty$*. Besides that,*

$$a_\infty(e^\sigma, e^\sigma w) = b(v,w) \qquad \forall v, w \in V'$$

and

$$e^\sigma v \in \mathbb{K}_\infty \Longleftrightarrow v \in \mathbb{K}'$$

Then

$$v_\infty \text{ is solution of (31) with } g_\infty \equiv 1 \Longleftrightarrow e^{-\sigma} v_\infty \text{ is solution of (36).} \tag{37}$$

Recalling that $s_n = \int_{q_s}^{1} \frac{g_n(\tau)}{\tau} d\tau$, since $g_n \longrightarrow g_\infty \equiv 1$ when $n \to +\infty$, uniformly on the compact subsets, then $s_n \longrightarrow s_\infty = -\log q_s$, when $n \to +\infty$.

Choose

$$s_0 \geq s_n, \qquad n \in I\!N \cup \{+\infty\},$$

and let

$$\Omega_{s_0} =]\theta_B, 0] \times]\mu, s_0[\cup]0, \theta_A[\times]\mu, +\infty[.$$

Given $v_n \in V_n(\Omega_{s_n})$, $n \in I\!N \cup \{+\infty\}$, extend v_n to Ω_{s_0} as follows:

$$\widehat{v}_n(\theta, \sigma) = \begin{cases} v_n(\theta, \sigma) & \text{if } \in \Omega_{s_n} \\ v_n(\theta, s_n) & \text{if } (\theta, \sigma) \in \Omega_{s_0} \backslash \Omega_{s_n}. \end{cases}$$

Notice that k_n is a continuous function defined in $[\mu, +\infty[$ and $\lim_{\sigma+\infty} k_n(\sigma)$ is finite and bounded above by a constant a_1 independent of n. On the other hand, $k_n > 0$ for all $n \in I\!N$. So we can say that

$$\exists a_0,\ a_1 > 0\ \forall n \in I\!N\ \forall \sigma \in [\mu, +\infty[: \quad a_0 \leq k_n(\sigma) \leq a_1. \tag{38}$$

Define now

$$W = \{v \in H'(\Omega_{s_0}) : v|_{\partial\Omega_{s_0} \backslash \Gamma_{s_0}} \equiv 0\} \tag{39}$$

Let

$$Z = e^{-a_0 \sigma} W, \qquad \|z\|_Z = \|e^{a_0 \sigma} z\|_W$$

Theorem 2.7 *Suppose that* g_n *verifies* (20), $n \in I\!N \cup \{+\infty\}$. *If* v_n *is the solution of problem* (31) *with density function* g_n *and* v_∞ *is the solution of the same problem with density function* $g_\infty \equiv 1$ *then*

$$\|\widehat{v}_n - \widehat{v}_\infty\|_Z \to 0 \qquad \textit{when } n \to +\infty. \tag{40}$$

Sketch of the proof:

- The bilinear form a_n is coercive in

$$\pi_n = \{v \in V_n(\Omega_{s_0}) : v(0, \sigma) = H \text{ if } \sigma > \sigma_n\}.$$

Notice that $\pi_n \supseteq \widehat{I\!K}_n$ and the constant of coerciveness (which we call α) can be choosen independently of $n \in I\!N \cup \{+\infty\}$.

- Recall that $\exists a_0, b_0 > 0 \forall \sigma \in [\mu, +\infty[: \quad a_0 \leq k_n(\sigma) \leq b_0$.

Notice that if $V_n = q_n^{-1} W$ we have

$$\exists C_1,\ C_2 > 0\ \forall v \in V_n \quad C_1 \|q_n v\|_{H^1} \leq \|v\|_{V_n} \leq C_2 \|q_n v\|_{H^1}$$

- Recall that $g_n \to 1$ uniformly on the compact subsets, when $n \to +\infty$.

Choose $C_n \geq 1$, $C_n \to 1$ when $n \to +\infty$, $g_n(0) \leq C_n$. Since g_n is decreasing, $g_n \leq C_n$. Let $\gamma_n = e^{-\sigma/C_n}$ and $j_n = \frac{1}{C_n^2}$. Let $Z_n = \gamma_n^{-1}W$. Then $Z_n \hookrightarrow V_n$ and $\|v\|_{V_n} \leq C\|v\|_{Z_n}$, C constant independent of n.

Let

$$b_n(u,v) = \int_\Omega e^{-2\sigma/C_n}(C_n^2 u_\sigma v_\sigma + u_\theta v_\theta - uv)d\theta d\sigma$$

and w_n be the solution of problem (31) with bilinear form b_n instead of a_n and convex set $\widehat{I\!M}_n$ defined as $I\!K_n$ but with function γ_n and j_n instead of q_n and k_n.

It is enough to prove that $\|w_n - \widehat{v}_n\|_Z \to 0$ and $\|w_n - \widehat{v}_\infty\|_Z \to 0$ when $n \to +\infty$.

• $\|v_n\|_{V_n}$ is bounded independently of n and

$$\begin{aligned}\|w_n - \widehat{v}_n\|_Z^2 &\leq \frac{1}{\alpha} a_n(w_n - \widehat{v}_n, w_n - \widehat{v}_n)\\ &= \frac{1}{\alpha}\left[-a_n(\widehat{v}_n, w_n - \widehat{v}_n) - a_n(w_n, \widehat{v}_n - w_n)\right]\end{aligned}$$

• Let $\Psi \in C^1([\theta_B, \theta_A])$ be such that $\Psi(\theta_B) = \Psi(\theta_A) = 0, \Psi(0) = 1$ and $\Psi \geq 0$.

• Notice that $W^{1,\infty} \cap W$ is dense in W.

Choose $\overline{w}_n \in q_n^{-1}W^{1,\infty} \cap V_n$ such that $\|w_n - \overline{w}_n\|_{V_n} \leq \frac{1}{n}$ and $\overline{w}(0,\sigma) = H$ if $\sigma \geq \sigma_n$. Define

$$\tilde{w}_n(\theta,\sigma) = \overline{w}_n(\theta,\sigma) + [\overline{w}_n(0,\sigma) \vee \eta_n(\sigma) - \overline{w}_n(0,\sigma)]\Psi(\theta).$$

Then

$$\tilde{w}_n \in \widehat{I\!K}_n \cap q_n^{-1}W^{1,\infty} \quad \text{and} \quad \|w_n - \tilde{w}_n\|_{V_n} \to 0, \text{ when } n \to +\infty,$$

and

$$-a_n(\widehat{v}_n, w_n - \widehat{v}_n) = -a_n(\widehat{v}_n, \tilde{w}_n - \widehat{v}_n) - a_n(\widehat{v}_n, w_n - \tilde{w}_n)$$

$$\leq \textstyle\int_\Omega [R(\theta) + \varphi_n] q_n^2(\widehat{v}_n - \tilde{w}_n) + \|\widehat{v}_n\|_{V_n}\|w_n - \tilde{w}_n\|,$$

where $\varphi_n(\theta,\sigma) = \frac{1}{q_n(\sigma)}\Psi_\sigma(\theta, s_n)\chi_{E_n}(\theta, s_n)$, $E_=]\theta_{s_n}, 0[\times]s_n, s[$ appear because of the extensions of the function v_n to Ω.

• Analogously, we construct $z_n \in \widehat{I\!M}_n \cap \tau_n^{-1}W^{1,\infty}$ such that $\|z_n - \widehat{v}_n\|_{V_n} \to 0$ when $n \to +\infty$ and

$$\begin{aligned}-a_n(w_n, \widehat{v}_n - w_n) &\leq \|w_n\|_{V_n}\|\widehat{v}_n - z_n\|_{V_n} + \\ &+ \int_{\Omega_{s_0}} [R(\theta) + \xi_n]\tau_n^2(z_n - w_n) + \\ &+ (a_n - b_n)(w_n, z_n - w_n).\end{aligned}$$

Then

$$\|w_n - \widehat{v}_n\|_Z^2 \leq O(1) + \int_{\Omega_{s_0}} q_n^2[R + \varphi_n](\tilde{w}_n - \widehat{v}_n)$$

$$+ \int_{\Omega_s} \tau_n^2[R + \xi_n](z_n - w_n)$$

$$= 0(1) + \int_{\Omega_{s_0}} \tau_n^2 R[(\tilde{w}_n - \widehat{w}_n) + (z_n - \widehat{v}_n)]$$

$$+ (a_n - b_n)(w_n, z_n - w_n)$$

$$+ \int_{\Omega_{s_0}} (q_n^2 - \tau_n^2) R((\tilde{w}_n - \widehat{w}_n) + \int_{\Omega_{s_0} \setminus \Omega_{s_n}} \Phi_n.$$

It is not difficult to prove (see [7]) that $|\int_{\Omega s \setminus \Omega_{s_n}} \Phi_n| \leq C(s - s_n)$, C constant independent of n (this uses the fact that v_n and w_n are uniformly bounded in $C^{1,\alpha}(\Omega_{s_n})$.

Since

$$\|\tilde{w}_n - w_n\|_{V_{s_0}} \longrightarrow 0, \quad q_n^2 - \tau_n^2 \longrightarrow 0, \qquad \text{when } n \to +\infty,$$

uniformly on the compact subsets of $[\mu, +\infty[$, we have that

$$\|w_n - \widehat{v}_n\|_Z^2 \leq F(n) + C(s_0 - s_n), \quad \text{where } F(n) \to 0 \text{ when } n \to +\infty.$$

- We can also prove that

$$\|w_n - \widehat{v}_\infty\|_Z \leq G(n) + C(s_0 - s_n), \quad \text{where } G(n) \to 0 \text{ when } n \to +\infty,$$

C constant independent of n (this case is not easier then the other one). So, given $n_0 \in I\!N$, for $n \geq n_0$, for all $s \geq \max\{s_\infty, s_n : n \geq n_0\}$, we have

$$\|\widehat{v}_n - v_\infty\|_Z \leq \|\widehat{v}_n - \widehat{v}_\infty\|_Z \leq F(n) + G(n) + C(s - s_n).$$

Since $s_n \longrightarrow s_\infty$ when $n \to +\infty$, fixing $\varepsilon > 0$

$$\exists p \in I\!N \ \forall n \geq p \quad |s_n - s_\infty| < \varepsilon.$$

Let $s = s_\infty + \varepsilon$. Then, for $n \geq p$, we have $\|\widehat{v}_n - v_\infty\|_Z \leq F(n) + G(n) + 2\varepsilon C$ and so,

$$\|\widehat{v}_n - v_\infty\|_Z \longrightarrow 0 \quad \text{when } n \to +\infty.$$

■

To study the behavior of the free boundaries, we need the following

Lemma 2.8 *Let θ_* be the angle betwen the wake and the profile on the point of intersection of both (for the fluid with density g_n), $n \in I\!N \cup \{+\infty\}$*

Then

$$\theta_*^n \longrightarrow \theta_*^\infty \quad \textit{when } n \to +\infty.$$

Proof: See [7]. ∎

Proposition 2.9

$\forall \delta > 0$ (small enough) $\qquad \|\ell_n - \ell_\infty\|_{C^{0,\alpha}(\theta_*^\infty+\delta,\theta_A-\delta)} \longrightarrow 0 \quad$ *when* $n \to +\infty$.

Proof:

• Let $\gamma \leq 1, \gamma \equiv 1$ in a neighbourhood of $(\mathcal{L}_n \cup \mathcal{L}) \cap]\theta_B + \delta/2, \theta_A - \delta/2[\times \mathbb{R}^+$ and such that supp $\gamma \subseteq \Omega_s \backslash \Sigma$, where $\Sigma = \{0\} \times [\inf_n\{\sigma_n\}, +\infty[$.

• If v_n is solution of problem (31) then

$$a_n(\gamma v_n, v - \gamma v_n) \geq \int_{\Omega_{s_n}} \xi_n(\theta,\sigma)(v - \gamma v_n), \qquad \forall v \in \mathbb{K},$$

where

$$\begin{aligned} \xi_n(\theta,\sigma) &= \gamma q_n^2[R(\theta) + \varphi_n(\theta,\sigma)] - \left(\gamma_\sigma \frac{q_n^2}{k_n}\right)_\sigma - \gamma_\sigma \frac{q_n^2}{k_n}(v_n)_\sigma \\ &- q_n^2 \gamma_{\theta\theta} v_n - 2q_n^2 \gamma_\theta (v_n)_\theta \end{aligned}$$

$$a(\gamma v_\infty, v - \gamma v_\infty) \geq \int_{\Omega_s} \xi(\theta,\sigma)(v - \gamma v_\infty), \qquad \forall v \in \mathbb{K},$$

where

$$\begin{aligned} \xi(\theta,\sigma) &= \gamma e^{-2\sigma} R(v) + (\gamma_\sigma e^{-2\sigma} v_\infty)_\sigma - \gamma_\sigma e^{-2\sigma}(v_\infty)_\sigma - e^{-2\sigma}\gamma_{\theta\theta} v_\sigma \\ &- 2e^{-2\sigma}\gamma_\theta (v_\infty)_\theta. \end{aligned}$$

• Let $K = \gamma^{-1}(1)$. Since K is a compact subset of Ω_{s_∞},

$$a = d(K, \mathbb{R}^2 \backslash \Omega_{s_\infty}) > 0.$$

Since $s_n \longrightarrow s_\infty$ when $n \to +\infty$,

$$\exists p \in \mathbb{N}\ \forall n \geq p \qquad s_n > s_\infty - a/2$$

and so,

$$n \geq p \Longrightarrow K \subseteq \Omega_{s_n}$$

Let

$$\beta = \min_{K,n}\{q_n^2|R(\theta)|, e^{-2\sigma}|R(\theta)|\}.$$

Simple calculations show us that

$$a(\gamma v_n, v - \gamma v_n) \leq \int_{\Omega_{s\infty}} \xi_n(v - \gamma v_n) - \int_{\Omega_{s\infty}} \tau_n(v - \gamma v_n) \qquad \forall v \in \mathbb{K},$$

where

$$\tau_n = \left[\left(\frac{q_n^2}{k_n} - e^{-2\sigma}\right)(\gamma v_n)_\sigma\right]_\sigma + (q_{n^2} - e^{-2\sigma})(\gamma v_n)_{\theta\theta} + (q_n^2 - e^{-2\sigma})\gamma v_n.$$

Notice that, in K, $(\xi_n+\tau_n)(\theta,\sigma) = q_n^2 R(\theta)(q_n^2 - e^{-2\sigma})v_n \leq -\frac{\beta}{2} < 0$ (notice that $q_n^2 R(\theta) \leq -\beta, q_n^2 \longrightarrow e^{-2\sigma}$, when $n \to +\infty$, uniformly in K and v_n is uniformly bounded and positive in K). Then, by known results,

$$\begin{aligned}\|\ell_n - \ell_\infty\|_{L^1(\theta_B+\delta,\theta_A-\delta)} &\leq \|\chi_n - \chi_\infty\|_{L^1(K)} \\ &\leq \frac{2}{\beta}\int_{\Omega_{s_\infty}} |(\xi_n + \tau_n) - \xi|\end{aligned}$$

where

$$\chi_n = \chi_{\{v_n=0\}} \text{ and } \chi_\infty = \chi_{\{v_\infty=0\}}.$$

But

$$\int_{\Omega_{s_\infty}} |\xi_n - \xi| \longrightarrow 0 \text{ and } \int_{\Omega_{s_\infty}} |\tau_n| \to 0 \text{ when } n \to +\infty,$$

since $q_n \longrightarrow e^{-\sigma}$, $q_n' \longrightarrow -e^{-\sigma}$, $k_n \longrightarrow 1$, $k_n^1 \to 0$, when $n \to +\infty$, uniformly on the compact subsets of $\mathbb{R}$ and $q_n v_n$ and $q_\infty v_\infty$ are bounded in $W^{2,p}_{loc}$ (independently of n)

Since ℓ_n are Lipschitz functions, using the Gagliardo-Niremberg inequality, we conclude that

$$\|\ell_n - \ell_\infty\|_{C^{0,\alpha}(\theta_B+\delta,\theta_A-\delta)} \longrightarrow 0 \qquad \text{when } n \to +\infty.$$

■

3 Continuous Dependence

We are going to prove in this section that the solution u of the variational inequality (18) depend continuously on the data, more precisely, on the geometry of the profile $\mathcal{P}$, with fixed height H (or equivalently on $R(\theta)$,on θ_A and θ_B), on the velocity at infinity q_∞, on the velocity at the wake q_s, on the density ρ and on the semi-height of the channel h. Many details are here ommited, since the idea of the proof follows the ideas of last section.

Let V_s be defined as in (14), for $\Omega_s =]\theta_B, 0] \times]\mu, s[\cup]0, \theta_A[\times]\mu, +\infty[$ and $V_{\widehat{s}}$ analogously, for $\Omega_{\widehat{s}} =]\theta_{\widehat{B}}, 0] \times]\mu, \widehat{s}[\cup]0, \theta_{\widehat{A}}[\times]\mu, +\infty[$. Given any function $v \in V_s$ (analogously $\widehat{v} \in V_{\widehat{s}}$), extend it to $\Omega =]\theta_B \wedge \theta_{\widehat{B}}, 0] \times]\mu, s \vee \widehat{s}[\cup [0, \theta_A \vee \theta_{\widehat{A}}[\times]\mu, +\infty[$, as follows:

$$\overline{v}(\theta,\sigma) = \begin{cases} v(\theta,\sigma) & \text{if } (\theta,\sigma) \in \Omega_s, \\ v(\theta,s) & \text{if } (\theta,\sigma) \in]\theta_B, \theta_A[\times [s, s \vee \widehat{s}[, \\ 0 & \text{if } \theta \leq \theta_B \vee \theta_{\widehat{B}} \text{ or } \theta \geq \theta_A \wedge \theta_{\widehat{A}}. \end{cases} \tag{41}$$

Remark 3.1 *We define*

$$V = \{v : \ (q \wedge \widehat{q})^2 v \in L^2(\Omega), \ (q \wedge \widehat{q})^2 v_\theta \in L^2(\Omega), (q \wedge \widehat{q})^2 \ v_\sigma \in L^2(\Omega), \\ v|_{\partial\Omega \setminus \Gamma} \equiv 0\},$$

where $\Gamma =]\theta_B \wedge \theta_{\widehat{B}}, 0[\times\{s \vee \widehat{s}\}$ *with the norm*

$$\|v\|_V = \int_\Omega (q \wedge \widehat{q})^2 (v_\sigma^2 + v_\theta^2 + v^2) d\theta d\sigma.$$

Notice that, since $\exists a_0 \ \forall \sigma \in [\mu, +\infty[: \quad k(\sigma) \geq a_0$, *if* $v \in V_s$, *then the extension* $\overline{v}$ *of* v *belongs to* V *and* $\|\overline{v}\|_V \leq C\|\overline{v}\|_{\overline{V}}$, *where* $\overline{V}$ *represents the space defined as* V_s *with the domains of definitions of its functions equal to* Ω.

We are going to ommit the bar over the extension of v, *since there is no risk of confusion. All the functions we are going to consider from now on are the extensions to* Ω *of functions defined in* Ω_s *or* $\Omega_{\widehat{s}}$. ∎

We define the thin obstacle η as before (and $\widehat{\eta}$),

$$\eta(\sigma) = H - hq_\infty g(q_\infty) \int_\sigma^{\sigma_\infty} \frac{k(\tau)}{q(\tau)} d\tau.$$

Theorem 3.2 *Let* u *and* $\widehat{u}$ *denote the solutions of problem* (18) *with data* R, θ_A, θ_B, q_∞, q_s, ρ, h *and* $\widehat{R}$, $\widehat{\theta_A}$, $\widehat{\theta_B}$, $\widehat{q_\infty}$, $\widehat{q}_s$, $\widehat{\rho}$, $\widehat{h}$, *respectively.*

Then we have

$$\|u - \widehat{u}\|_V^2 C_1 \|(\eta - \widehat{\eta})\|_V^2 + C_2|s - \widehat{s}| + C_3\|q - \widehat{q}\|_{L^1(\Omega)}^2 + C_4\|q^2 R - \widehat{q}^2 \widehat{R}\|_{L^1(\Omega)} \quad (42)$$

where C_1, C_2, C_3 *and* C_4 *are constants independent of the given data.*

Remark 3.3 *We would like to remark that* $\|u - \widehat{u}\|_V^2$ *depend on* θ_A, $\widehat{\theta}_A$, θ_B, $\widehat{\theta}_B$, *through* R *and* $\widehat{R}$. *Recall that* $q = g^{-1}(\rho)$ *and* $\widehat{q} = \widehat{g}^{-1}(\widehat{\rho})$. ∎

Proof: Extend u and $\widehat{u}$ to Ω as it was done before. Then calling to the extensions still u and $\widehat{u}$ and calling $a_q(u, v) = \int_\Omega q^2 \left(\frac{1}{k} u_\sigma v_\sigma + u_\theta v_\theta - uv\right)$ in order to avoid confusion, we easily see that

$$\begin{cases} u \in I\!\!K \\ a_q(u, v - u) \geq \int_\Omega q^2(\sigma) \left[R + \varphi\right](\sigma)(v - u) d\theta d\sigma, \quad \forall v \in I\!\!K, \end{cases} \quad (43)$$

where $\varphi(\theta, \sigma) = \frac{1}{q(\sigma)} \Psi_\sigma(\theta, s) \chi_E$, $E =]\theta_s, 0[\times]s, s \vee \widehat{s}[$ (with $\int_{\theta_B}^0 |\Psi_\sigma(\theta, s)| d\theta \leq C$, C constant independent of s), $I\!\!K = \{v \in V : \ v \geq 0, \ v(0, \sigma) = H$ if $\sigma \geq$

σ_∞, $v(0,\sigma) \geq \eta(\sigma)$ if $\sigma < \sigma_\infty\}$ and $\widehat{u}$ is the solution of the same variational inequality with h replaced by $\widehat{h}$, R by $\widehat{R}$, σ_∞ by $\widehat{\sigma}_\infty$, η by $\widehat{\eta}$, k by $\widehat{k}$ and q by $\widehat{q}$.

Notice that u may not belong to $\widehat{I\!K}$ as well as $\widehat{u}$ may not belong to $I\!K$. Let $u_n \in I\!K \cap W^{1,\infty}(\Omega)$ be such that $\|u - u_n\|_V \leq \frac{1}{n}$ and define

$$\widehat{v}_n(\theta,\sigma) = u_n(\theta,\sigma) + (\widehat{\eta} \vee u_n(0,\sigma) - u_n(0,\sigma)\,\varphi(\theta),$$

with $\varphi \in C^1([\theta_B, \theta_A])$, $\varphi(\theta_B) = \varphi(\theta_A) = 0$, $\varphi(0) = 1$; we have then that $\widehat{v}_n \in \widehat{I\!K}$ (notice that $v_n \in V_{\widehat{s}}$). Analogously, choosing $\widehat{u}_n \in \widehat{I\!K} \cap W^{1,\infty}(\Omega)$ such that $\|\widehat{u} - \widehat{u}_n\|_V \leq \frac{1}{n}$, define

$$v_n(\theta,\sigma) = \widehat{u}_n(\theta,\sigma) + (\eta \vee \widehat{u}_n(0,\sigma) - \widehat{u}_n(0,\sigma))\varphi(\theta);$$

then $v_n \in I\!K$.

We have then that

$$\begin{aligned} \|v_n - u\|_V &\leq \|v_n - u_n\|_V + \|u_n - u\|_V \\ &\leq \frac{1}{n} + \|(\widehat{\eta} \vee u_n(0,\sigma) - u_n(0,\sigma))\varphi(\theta)\|_V, \end{aligned}$$

and since

$$\widehat{\eta} \vee u_n(0,\sigma) - u_n(0,\sigma) \begin{cases} = 0 & \text{if } u_n(0,\sigma) \geq \widehat{\eta}, \\ \leq \widehat{\eta} - \eta & \text{if } u_n(0,\sigma) \leq \widehat{\eta}, \end{cases}$$

we have $0 \leq \widehat{\eta} \vee u_n(0,\sigma) - u_n(0,\sigma) \leq |\widehat{\eta} - \eta|$ and so

$$\begin{aligned} \|v_n - u\|_V &\leq \frac{1}{n} + C\|(\widehat{\eta} \vee u_n(0,\sigma) - u_n(0,\sigma))\varphi(\theta)\|_V \\ &\leq \frac{1}{n} + C\|\varphi(\eta - \widehat{\eta})\|_V, \end{aligned}$$

where C is a constant independent of the given data. The estimation of $\|\widehat{v}_n - \widehat{u}\|_V$ is obtained in the same way.

So, if α is the constant of coerciveness of a, and since $a(v,w) \leq \|v\|_V \|w\|_V$, we have

$$\begin{aligned} \alpha\|u - \widehat{u}\|_V^2 &\leq a(u - \widehat{u}, u - \widehat{u}) \\ &= -a(u, \widehat{u} - u) - a(\widehat{u}, u - \widehat{u}) \\ &\leq -a(u, v_n - u) - a(u, \widehat{u} - v_n) - a(\widehat{u}, \widehat{v}_n - \widehat{u}) - a(\widehat{u}, \widehat{v}_n - u) \\ &= -(a - a_q)(u, v_n - u) - a_q(u, v_n - u) - a(u, \widehat{u} - v_n) \\ &\quad - \left(a - a_{\widehat{q}}\right)(\widehat{u}, \widehat{v}_n - \widehat{u}) - a_{\widehat{q}}(\widehat{u}, \widehat{v}_n - \widehat{u}) - a(\widehat{u}, \widehat{v}_n - u) \end{aligned}$$

$$\leq \; C\int_\Omega (q\wedge\widehat{q}-\widehat{q})^2+\int_\Omega\left[q^2R(\theta)+\psi\right](v_n-u)+\|u\|_V\|\widehat{u}-v_n\|_V$$

$$+C\int_\Omega (q\wedge\widehat{q}-\widehat{q})^2+\int_\Omega\left[\widehat{q}^2\widehat{R}(\theta)+\widehat{\psi}\right](\widehat{v}_n-\widehat{u})+\|\widehat{u}\|_V\|\widehat{v}_n-u\|_V$$

$$\leq \; C_0\tfrac{1}{n}+C_1\|(\eta-\widehat{\eta})\|_V+C_2|s\vee\widehat{s}-s\wedge\widehat{s}|$$

$$+\int_\Omega\left(q^2R-\widehat{q}^2\widehat{R}\right)(u-\widehat{u})$$

$$\leq \; C_0\tfrac{1}{n}+C_1\|(\eta-\widehat{\eta})\|_V+C_2|s-\widehat{s}|+C_3\|q-\widehat{q}\|^2_{L^1(\Omega)}$$

$$+2H\|q^2R-\widehat{q}^2\widehat{R}\|_{L^1(\Omega)}$$

noticing that $\int_\Omega(-\psi)q^2(v_n-u)\leq C|s-\widehat{s}|$, as well as $\int_\Omega(-\widehat{\psi})q^2(v_n-\widehat{u})\leq C|s-\widehat{s}|$ and $|u-\widehat{u}|\leq 2H$.

So

$$\|u-\widehat{u}\|^2_V\leq C_1\|(\eta-\widehat{\eta})\|^2_V+C_2|s-\widehat{s}|+C_3\|q-\widehat{q}\|^2_{L^1(\Omega)}+C_4\|q^2R-\widehat{q}^2\widehat{R}\|_{L^1(\Omega)}.$$

■

Remark 3.4 *We would like to remark that, if u denotes the solution of the problem of a compressible fluid with density ρ and $\widehat{u}$ the solution of a problem of a compressible fluid with density $\widehat{\rho}$, with the other data fixed, we have, since $\rho=g(q)$ and $\widehat{\rho}=\widehat{g}(\widehat{q})$,*

$$\|u-\widehat{u}\|_V\leq C\|q-\widehat{q}\|^2_{L^1(\Omega)}=C\|g^{-1}(\rho)-\widehat{g}^{-1}(\widehat{\rho})\|^2_{L^1(\Omega)}.$$

As a particular situation, if u_∞ is the solution of the problem for an incompressible fluid, we have

$$\|u-u_\infty\|^2_V\leq C\|g^{-1}(\rho)-e^{-2\sigma}\|^2_{L^1(\Omega)}.$$

■

References

[1] Brézis, H., *A new method in the study of subsonic flows*, Lect. Notes Math. (Springer) (1975) 50-64.

[2] Brézis, H. & Duvaut, G., *Écoulements avec sillages autour d'un profil symetrique sans incidence*, C. R. Acad. Sc. Paris **276** (1973) 975-993.

[3] Brézis, H. & Stampacchia, G., *The hodograph method in fluid-dynamics in the light of variational inequalities*, Arch. Rat. Mech. and Anal. **61** (1976) 1-18.

[4] Díaz, J. I., *Tecnica de supersoluciones locales para problemas estacionarios no lineales. Aplicacion al estudio de flujos subsonicos*, Real Acad. Cien. Madrid **XVI** (1982).

[5] Hummel, R., *The hodograph method for convex profiles*, Ann. Scuo. Norm. Sup. Pisa **9** IV (1982) 341-363.

[6] Santos, L., *Variational convergences of a flow with a wake in a channel past a profile*, Bollettino U.M.I. **7** 2-B (1988) 109-125.

[7] Santos, L., *On the variational inequality approach to compressible flows via hodograph transformation*, Rev. Mat. Univ. Comp. Madrid **6** 2 (1993) 333-374.

[8] Shimborski, E., *Variational methods applied to the study of symmetric flows in laval nozzles*, Commun. P. D. E. **4** (1979) 41-77.

[9] Shimborski, E., *Variational inequalities arising in the theory of two dimensional potential flows*, Nonlinear Analysis T. M. A. **5** (1981) 434-444.

[10] Tomarelli, F. , *Hodograph method and variational inequalities in fluid-dynamics*, Proocedings of "Free Boundary Problems", Pavia, 1979.

STABILITY THRESHOLDS FOR CONVECTION WHEN THE VISCOSITY HAS A GENERAL FORM OF TEMPERATURE DEPENDENCE

Brian STRAUGHAN
Department of Mathematics,
The University,
Glasgow G12 8QW,
Scotland.

Abstract

A nonlinear energy stability analysis is presented for convection when the viscosity may have a general dependence on temperature. Especial attention and numerical calculations are given when the viscosity has the form advocated by Tippelskirch [16].

1. Introduction.

Much attention has recently been devoted to the subject of thermal convection when the variation of viscosity with temperature is accounted for, see e.g. [1-4], [6-10], [12], [15-17], and see also several other references given in [14]. Since the viscosity may change dramatically with temperature, e.g. page 6-158 of Lide [5] shows the dynamic viscosity of glycerin drops from 6.71×10^6 cp. at $-42°$C to 629 cp. at 30°C, there are many fluids for which it may be argued that thermal convection studies must account for viscosity variation; while Glycerin exhibits a dramatic viscosity change there are many other examples, particularly oily fluids, quoted in the tables in Lide [5].

The object here is to present full details of a nonlinear energy stability analysis for convection when the viscosity may have an arbitrary temperature dependence, although we concentrate on the relationship advocated by Tippelskirch [16]. For completeness we outline previous analyses which employ energy stability theory for the temperature dependent viscosity convection problem. Straughan [12] utilized one of the the viscosities of Torrance & Turcotte [17], namely that which selects for the kinematic viscosity ν,

$$\nu = \nu_0 e^{-\gamma z}, \tag{1.1}$$

ν_0, γ constants; in (1.1) z is the vertical coordinate pointing upward through the fluid layer. This was extended by Richardson & Straughan [8] who were able to

S. N. Antontsev et al. (eds.), Energy Methods in Continuum Mechanics, 145–157.

handle the more realistic viscosity of Palm *et al.* [7], namely, with T denoting temperature,

$$\nu(T) = \nu_0\big(1 - \gamma(T - T_0)\big), \tag{1.2}$$

γ, T_0 constants. Franchi & Straughan [3] also utilize (1.2) but allow the fluid to be one of generalised second grade type; their analysis is able to employ the richer structure provided by the more exotic equations to obtain stronger mathematical energy stability. Richardson & Straughan [9] also use (1.2) but in a porous medium where they find need also for a Brinkman equation of momentum. A departure from the linear relation (1.2) was developed by Capone & Gentile [1] who choose

$$\nu(T) = \nu_0 e^{-\gamma(T-T_0)}; \tag{1.3}$$

this is a realistic fit for many real fluids as pointed out by e.g. Torrance & Turcotte [17].

The object of the current paper is primarily to focus attention on the viscosity relation suggested by Tippelskirch [16] in his fundamental paper in which he infers by his experiments on liquid sulphur that the viscosity dependence actually determines whether the fluid rises or descends in the centre of the convection cell. The fluid viscosity advocated by Tippelskirch [16] is

$$\nu(T) = \frac{C}{1 + AT + BT^2}, \tag{1.4}$$

A, B, C constants (in this work they are positive). To give specific examples of curve fits for (1.4) we may use data of Lide [5] to find for:

(A) *Aniline* ($C_6H_5NH_2$.)

Temperature (°C)	ν (cm^2 sec^{-1})
-6	0.42593
0	0.31482
5	0.24877
10	0.20062

Table 1. Kinematic Viscosity vs. Temperature for Aniline.

This data may be fitted to (1.4) to find (using only the first three data points)

$$\nu(T) = \frac{0.31482}{1 + 0.48727 \times 10^{-1}T + 0.87490 \times 10^{-3}T^2}. \tag{1.5}$$

(B) *Nitrobenzene* ($C_6H_5NO_2$.)

Temperature (°C)	ν (cm^2 sec^{-1})
2.95	2.4290
9.92	2.0701
14.94	1.8698

Table 2. Kinematic Viscosity vs. Temperature for Nitrobenzene.

The data of table 2 leads to the curve fit

$$\nu(T) = \frac{2.6202}{1 + 0.26641 \times 10^{-1}T + 0.14832 \times 10^{-4}T^2} \tag{1.6}$$

Numerical results specific to (1.5) and (1.6) are included in section 3.

2. Nonlinear energy stability analysis.

To develop a nonlinear energy stability analysis let us begin with a layer of heat conducting viscous fluid contained between the planes $z = 0$ and $z = d$ with gravity acting in the negative $z-$direction. The relevant equations of momentum, continuity, and energy balance with ν dependent on temperature T, are

$$\begin{aligned} v_{i,t} + v_j v_{i,j} =& -\frac{1}{\rho}p_{,i} + 2\{\nu(T)D_{ij}\}_{,j} + g\delta_{i3}\alpha T, &(2.1)\\ v_{i,i} =&0, &(2.2)\\ T_{,t} + v_i T_{,i} =&\kappa\Delta T, &(2.3)\end{aligned}$$

where $g, \alpha, \kappa, p, v_i, T, \rho$ are gravity, thermal expansion coefficient, thermal diffusivity, pressure, velocity, temperature, and (constant) density, respectively; the usual Boussinesq approximation has been employed in the buoyancy term. The symbol Δ is the three-dimensional Laplacian, standard indicial notation is employed, and the tensor D_{ij} is the symmetric part of the velocity gradient, namely,

$$D_{ij} = \frac{1}{2}(v_{i,j} + v_{j,i}).$$

The boundaries $z = 0, d$ are assumed to be held at fixed, constant temperatures

$$T = T_1, \quad \text{at } z = 0, \qquad T = T_2, \quad \text{at } z = d, \qquad \text{where } T_1 > T_2,$$

and then the steady solution whose stability is under investigation is

$$\bar{v}_i \equiv 0, \qquad \bar{T} = -\beta z + T_1, \tag{2.4}$$

where β is the temperature gradient given by

$$\beta = \frac{\Delta T}{d}, \qquad \Delta T = T_1 - T_2(>0),$$

and the stationary solution is completed by calculating the steady pressure field $\bar{p}(z)$ from equation (2.1).

Perturbation variables (u_i, θ, π) to the velocity, temperature, and pressure fields are introduced and then with the non-dimensional variables suggested by the work of Richardson & Straughan [8] the non-dimensional perturbation equations become

$$u_{i,t} + u_j u_{i,j} = -\pi_{,i} + 2\{\nu(\bar{T}+\theta)d_{ij}\}_{,j} + R\theta\delta_{i3}, \tag{2.5}$$

$$u_{i,i} = 0, \tag{2.6}$$

$$Pr(\theta_{,t} + u_i\theta_{,i}) = Rw + \Delta\theta, \tag{2.7}$$

where $Ra = R^2$ is the Rayleigh number, the spatial domain has become

$$\{(x,y) \in \mathbf{R}^2\} \times \{z \in (0,1)\},$$

$\nu(\bar{T}+\theta)$ is the non-dimensional form of viscosity (which arises from

$$\nu(\bar{T}+\theta) = \frac{1}{1 + A(\bar{T}+\theta) + B(\bar{T}+\theta)^2},$$

in the case of the Tippelskirch [16] representation), and

$$d_{ij} = \frac{1}{2}(u_{i,j} + u_{j,i}).$$

In addition to the constant temperature boundary conditions we suppose the bounding surfaces are free from tangential stress so that the boundary conditions on the perturbation velocity field (u, v, w) and temperature θ are

$$w = \frac{\partial^2 w}{\partial z^2} = \Delta w = 0, \quad \frac{\partial u}{\partial z} = \frac{\partial v}{\partial z} = 0; \quad \theta = \Delta\theta = 0; \quad z = 0, 1. \tag{2.8}$$

Convection cells typically have an (x, y) periodic spatial behaviour and so let V denote a period cell for the perturbation variables. Denote by $\|\cdot\|$ and $<\cdot>$ the norm on $L^2(V)$ and integration over V, respectively. The nonlinear energy stability result we derive begins by multiplying (2.5) by u_i, and integrating the result over V to find

$$\frac{1}{2}\frac{d}{dt}\|\mathbf{u}\|^2 = R < \theta w > -2 < \nu(\bar{T}+\theta)d_{ij}d_{ij} >, \tag{2.9}$$

and then we multiply (2.7) by θ, and integrate the result over V to obtain

$$\frac{1}{2}Pr\frac{d}{dt}\|\theta\|^2 = R < w\theta > -\|\nabla\theta\|^2. \tag{2.10}$$

Next, the viscosity is expanded in a Taylor series

$$\nu(\bar{T}+\theta)=\nu(\bar{T})+\theta\nu'(\hat{T}), \qquad (2.11)$$

where $\hat{T}\in(\bar{T},\bar{T}+\theta)$, (or $\hat{T}\in(\bar{T}+\theta,\bar{T})$ if $\theta<0$,) and (2.9) becomes

$$\frac{1}{2}\frac{d}{dt}\|\mathbf{u}\|^2 = R<\theta w> -2<F(z)d_{ij}d_{ij}> +2<\nu'(\hat{T})\theta d_{ij}d_{ij}>, \qquad (2.12)$$

where we have put $F(z)=\nu(\bar{T})$.

Equations (2.10) and (2.12) are added together to obtain a basic energy identity of form

$$\frac{dE}{dt}=RI-D+N, \qquad (2.13)$$

where E is a natural kinetic - like energy, I is the quadratic production term, D is the quadratic dissipation, and N contains the cubic nonlinearities; these terms are now written in detail:

$$E=\frac{1}{2}\|\mathbf{u}\|^2+\frac{1}{2}Pr\|\theta\|^2, \qquad (2.14)$$

$$I=2<\theta w>, \qquad (2.15)$$

$$D=2<F(z)d_{ij}d_{ij}>+\|\nabla\theta\|^2, \qquad (2.16)$$

$$N=2<\nu'(\hat{T})\theta d_{ij}d_{ij}>. \qquad (2.17)$$

The next step is to define the variational critical energy stability number R_E by

$$\frac{1}{R_E}=\max_{\mathcal{H}}\frac{I}{D}, \qquad (2.18)$$

where $\mathcal{H}$ is the space of admissible solutions we are working with. The nonlinear energy stability threshold requires

$$R<R_E, \qquad (2.19)$$

and so we define the number a by

$$a=\frac{R_E-R}{R_E}(>0), \qquad (2.20)$$

and from (2.13) we may derive

$$\frac{dE}{dt}\le -aD+N. \qquad (2.21)$$

The difficulty in proceeding beyond (2.21) is that the terms in D are evidently not sufficient to control the nonlinearity of N. The procedure adopted in [8] is to introduce extra terms in the "energy" and work with a generalised energy, or Lyapunov functional. To this end multiply (2.5) by Δu_i and integrate over V to obtain

$$\begin{aligned}\frac{1}{2}\frac{d}{dt}\|\nabla\mathbf{u}\|^2 =& - R < \Delta w\theta > - < \nu'(\bar{T}+\theta)\bar{T}_{,j}(u_{i,j}+u_{j,i})\Delta u_i > \\ & - 2 < \nu(\bar{T}+\theta)\Delta u_i\Delta u_i > + < \Delta u_i u_j u_{i,j} > \\ & - 2 < \nu'(\bar{T}+\theta)\Delta u_i\theta_{,j}d_{ij} >, \\ =& Q_1 - D_1 + N_1, \end{aligned} \tag{2.22}$$

where Q_1, D_1 and N_1 are, respectively, a quadratic production term, a quadratic dissipation term, and a term involving cubic nonlinearities, defined specifically by

$$\begin{aligned} Q_1 =& - R < \Delta w\theta > - < \nu'(\bar{T}+\theta)\bar{T}_{,j}u_{i,j}\Delta u_i > \\ & - < \nu'(\bar{T}+\theta)\bar{T}_{,j}u_{j,i}\Delta u_i >, \end{aligned} \tag{2.23}$$

$$D_1 = 2 < \nu(\bar{T})\Delta u_i\Delta u_i >, \tag{2.24}$$

$$\begin{aligned} N_1 =& < \Delta u_i u_j u_{i,j} > -2 < \nu'(\bar{T}+\theta)\Delta u_i\theta_{,j}d_{ij} > \\ & - 2 < \nu'(\hat{T})\theta\Delta u_i\Delta u_i > . \end{aligned} \tag{2.25}$$

In deriving (2.22), (2.11) has been employed.

Next take Δ of (2.7), multiply the result by $\Delta\theta$ and then integrate over V to see that

$$\frac{1}{2}Pr\frac{d}{dt}\|\Delta\theta\|^2 = Q_2 - D_2 + N_2, \tag{2.26}$$

where Q_2, D_2, N_2 have similar meanings to Q_1, D_1, N_1, and are given by

$$Q_2 = R < \Delta w\Delta\theta >, \tag{2.27}$$

$$D_2 = \|\nabla\Delta\theta\|^2, \tag{2.28}$$

$$N_2 = 2Pr < u_{i,j}\theta_{,j}\Delta\theta_{,i} > -2Pr < \Delta\theta\Delta u_i\theta_{,i} > . \tag{2.29}$$

The tricky part is in arranging equations (2.22) and (2.26) in the manner above. From this point the analysis proceeds according to the prescription given in [8], hence only brief details are given.

A generalised energy $\mathcal{E}(t)$ is defined by

$$\mathcal{E}(t) = E(t) + \frac{1}{2}\eta\|\nabla\mathbf{u}\|^2 + \frac{1}{2}\zeta Pr\|\Delta\theta\|^2, \tag{2.30}$$

where η, ζ are positive coupling parameters. An appropriate combination of (2.21), (2.22) and (2.26) yields our generalised energy inequality

$$\frac{d\mathcal{E}}{dt} \le -\mathcal{D} + \mathcal{Q} + \mathcal{N}, \tag{2.31}$$

where the total dissipation $\mathcal{D}$ is defined by

$$\mathcal{D} = aD + \eta D_1 + \zeta D_2, \tag{2.32}$$

Q contains quadratic terms and is

$$Q = \eta Q_1 + \zeta Q_2, \tag{2.33}$$

and finally $\mathcal{N}$ represents the cubic nonlinear terms, namely

$$\mathcal{N} = N + \eta N_1 + \zeta N_2. \tag{2.34}$$

The coupling parameters η, ζ are chosen in such a way that

$$Q \leq \frac{1}{2}\mathcal{D},$$

and then by use of inequalities such as

$$\sup_{\mathcal{H}} |\nabla\theta| \leq c\|\nabla\Delta\theta\|,$$

$\mathcal{N}$ is bounded in the sense that

$$\mathcal{N} \leq c_1 \mathcal{D}\mathcal{E}^{1/2}. \tag{2.35}$$

The upshot of these estimates is that from (2.31) we arrive at an inequality of form

$$\frac{d\mathcal{E}}{dt} \leq -\frac{1}{2}\mathcal{D} + A\mathcal{E}^{1/2}\mathcal{D}, \tag{2.36}$$

where A is a constant which is completely determinable. The details leading to (2.35) and the calculation of the constant c_1 follow from similar calculations in Richardson & Straughan [8], although to arrive at the bound (2.35) we need to work with a viscosity function $\nu(T)$ such that

$$|\nu'(T)| \leq K, \tag{2.37}$$

for some constant K and $\forall T \in [\bar{T}, \bar{T}+\theta]$. For the relation (1.4) of Tippelskirch [16] it is easy to show (2.37) is true. It would also be true for a generalised Tippelskirch viscosity of form

$$\nu = \frac{A_{n+1}}{1 + A_1 T + A_2 T^2 + \cdots + A_n T^n},$$

provided the A_i satisfy appropriate inequalities (see also the remark at the end of section 3).

From Poincaré's inequality $\exists$ a constant ξ such that $\mathcal{D} \geq \xi\mathcal{E}$ and it is then possible to show, cf. Straughan [13], that provided

$$\mathcal{E}^{1/2}(0) < \frac{1}{2A} \qquad \text{and} \qquad R < R_E, \tag{2.38}$$

then $\mathcal{E}(t) \to 0$ as $t \to \infty$ and nonlinear stability is established.

In the next section we analyse the implications of (2.38) for the viscosity relation (1.4).

3. Numerical results for the Tippelskirch viscosity.

The stability obtained in section 2 is conditional in the sense that the initial energy is restricted by condition $(2.38)_1$. We now concentrate on calculating the critical Rayleigh number for nonlinear energy stability, R_E.

The eigenvalue problem which arises from the variational characterisation of the nonlinear stability threshold R_E, (2.18), is the same as the one which is found in the problem of the determination of the linear stability boundary. From (2.5) - (2.7) this is seen to be

$$\left.\begin{aligned} &\{\nu(\bar{T})(u_{i,j}+u_{j,i})\}_{,j} + R_E\theta\delta_{i3} = \pi_{,i},\\ &u_{i,i}=0,\\ &R_E w + \Delta\theta = 0, \end{aligned}\right\} \tag{3.1}$$

In (3.1) π is a Lagrange multiplier and we solve this system on

$$z \in (0,1), \quad (x,y) \in \mathbf{R}^2,$$

for the smallest eigenvalue R_E. For our numerical calculations we restrict attention to the Tippelskirch [16] relation (1.4) for the viscosity, and for this

$$\begin{aligned} F(z) =& \nu(\bar{T})\\ =& \frac{1}{1+(\xi_1-\Gamma_1 z)+(\xi_2-\Gamma_2 z)^2}, \end{aligned} \tag{3.2}$$

where

$$\xi_1 = AT_1, \qquad \xi_2 = \sqrt{B}T_1, \tag{3.3}$$

and

$$\Gamma_1 = A\Delta T, \qquad \Gamma_2 = \sqrt{B}\Delta T. \tag{3.4}$$

To solve (3.1), π, u, v are eliminated by taking curlcurl of $(3.1)_1$ and then the third component of the resulting equation together with $(3.1)_3$ yields the system

$$\left.\begin{aligned} &F''\Big(\Delta^* w - \frac{\partial^2 w}{\partial z^2}\Big) - F\Delta^2 w - 2F'\Delta\frac{\partial w}{\partial z} - R_E\Delta^*\theta = 0,\\ &\Delta\theta + R_E w = 0, \end{aligned}\right\} \tag{3.5}$$

where Δ^* is the horizontal Laplacian, i.e.

$$\Delta^* = \frac{\partial^2}{\partial x^2} + \frac{\partial^2}{\partial y^2}.$$

A plane tiling spatial planform is assumed for w and θ, i.e. we write

$$w = W(z)f(x,y), \qquad \theta = \Theta(z)f(x,y),$$

where

$$\Delta^* f = -a^2 f,$$

a being the horizontal wavenumber; typically f represents a hexagonal pattern. Then system (3.5) becomes, with $D = d/dz$,

$$\left.\begin{aligned} D^4 W =& -2\frac{F'}{F}D^3W + \left(2a^2 - \frac{F''}{F}\right)D^2W + 2\frac{F'}{F}a^2DW \\ & - a^2\left(a^2 + \frac{F''}{F}\right)W + \frac{R_E a^2}{F}\Theta, \\ D^2\Theta =& a^2\Theta - R_E W, \end{aligned}\right\} \tag{3.6}$$

$z \in (0,1)$. We solve this for fixed a^2 for the smallest eigenvalue R_E, by using the compound matrix method, cf. Straughan [13], appendix 2, and we then find the critical Rayleigh number of energy stability as

$$Ra_0 = \min_{a^2} R_E^2(a^2). \tag{3.7}$$

The critical Rayleigh number Ra_0 as defined by (3.7) corresponds to

$$Ra_0 = \frac{\alpha g d^3 \Delta T}{\kappa \nu_0}, \tag{3.8}$$

where ν_0 is the viscosity of (1.4) evaluated at $T = 0°$C. There is some concern in the literature, cf. e.g. [18], as to what is the most useful Rayleigh number to employ in convection studies when the viscosity is temperature dependent. Other possibilities include those with the viscosity defined as its value at the lower (warmer) boundary, ν_{T_1}, or several writers advocate an average viscosity. Even this can be defined in several ways, for example with

$$\nu_m = \frac{1}{2}\big[\nu(T_1) + \nu(T_2)\big], \tag{3.9}$$

$$\nu_{T_m} = \nu\left(\frac{T_1 + T_2}{2}\right), \tag{3.10}$$

or

$$\nu_{av} = \frac{1}{d}\int_0^d \nu(\bar{T}(z))\,dz. \tag{3.11}$$

In the computational output presented below we give the critical values for Ra_0 as defined by (3.8), and we also give values for the Rayleigh number which utilizes (3.9), that is

$$Ra = \frac{\alpha g d^3 \Delta T}{\kappa \nu_m}. \tag{3.12}$$

In the computations we begin with a general study of the effect of the variation of ν with temperature on the Rayleigh number. It is seen that for $\xi_1 = \xi_2 = \Gamma_1 = \Gamma_2 = 0$ the critical Rayleigh number is the same as the classical one for two free surfaces. As the viscosity variation is increased the results in table 3 show a general decrease in Ra_m which may be interpreted as showing the decrease in viscosity makes convective fluid motion easier and the stability threshold is thus lowered.

ξ_1	ξ_2	Γ_1	Γ_2	Ra_0	Ra	a_c^2
0.0	0.0	0.0	0.0	657.5	657.5	4.935
0.1	0.05	0.02	0.01	602.2	657.6	4.833
0.2	0.1	0.04	0.02	553.5	657.5	4.833
0.1	0.02	0.02	0.004	603.1	657.6	4.833
0.2	0.04	0.04	0.008	556.7	657.5	4.833
0.1	0.05	0.1	0.05	626.3	656.8	4.833
0.2	0.1	0.2	0.1	597.7	654.5	4.833
0.1	0.02	0.1	0.02	626.7	656.7	4.833
0.2	0.04	0.2	0.04	599.3	654.2	4.833
1.0	0.5	0.5	0.25	350.5	646.4	4.833

Table 3. Critical Rayleigh number of energy stability vs. critical wavenumber.

For completeness we include calculations for aniline and nitrobenzene when $T_1 = 10°\mathrm{C}$ and $\Delta T = 2°\mathrm{C}$. The viscosity fits for these fluids are given, respectively, by (1.5) and (1.6). The two surfaces are assumed *fixed* in this case.

Fluid	Ra	Ra_0	a_c^2
Aniline	1706.1	1131.7	9.721
Nitrobenzene	1707.2	1376.3	9.721

Table 4. Critical Rayleigh and wave numbers for Aniline and Nitrobenzene.

By using the Ra_0 values in table 4 we may fix the temperature difference and find the critical depth. For instance, for nitrobenzene data of Stiles [11] yields

$$\alpha = 8.3 \times 10^{-4}\,(°\mathrm{C}^{-1}),$$
$$\kappa = 0.807 \times 10^{-7}\ \mathrm{m}^2\,\mathrm{s}^{-1},$$

and then we take

$$g = 9.81\,\mathrm{m\,s^{-2}},$$
$$\nu_0 = 2.6202\ \mathrm{cm^2\,s^{-1}}.$$

Upon using the above data with (3.8) and the appropriate Ra_0 value from table 4 we find

$$d^3 \Delta T = 3.5742 \times 10^{-6},$$

and since $\Delta T = 2°\mathrm{C}$ we obtain

$$d_{\mathrm{crit}} = 1.213\,\mathrm{cm}.$$

Remarks. While we have concentrated here on relation (1.4) advocated by Tippelskirch [16] we could easily have employed other viscosity forms; all we require is $|\nu'(\hat{T})|$ bounded. For example, a superexponential relationship

$$\nu = K \exp\left(\frac{1}{A + BT + CT^2}\right), \tag{3.13}$$

could be employed for suitable A, B, C, with $C > 0$: this relation is employed by White [18] in his work with Golden Syrup, although there C is negative. Using data from table 1 we have fitted aniline by taking

$$K = 0.32598 \times 10^{-1}\ (\mathrm{cm^2\ sec^{-1}}),$$
$$A = 0.440966,$$
$$B = 0.950292 \times 10^{-2}\ (°\mathrm{C}^{-1}),$$
$$C = 0.143142 \times 10^{-3}\ (°\mathrm{C}^{-2}).$$

In addition, we could have allowed a nonlinear density relationship, such as

$$\rho = \frac{\rho_0}{1 + B_1 T + B_2 T^2 + B_3 T^3 + B_4 T^4}\,; \tag{3.14}$$

such densities have been employed in the literature in other convection analyses, see e.g. the account and references in Straughan [14], pages 81 - 84. For a relation like (3.14) it is necessary to expand the density in a manner similar to (2.11), except that here we need

$$\rho(\bar{T} + \theta) = \rho(\bar{T}) + \theta \rho'(\bar{T}) + \frac{1}{2}\theta^2 \rho''(\tilde{T}), \tag{3.15}$$

for some $\tilde{T} \in (\tilde{T}, \tilde{T} + \theta)$, and we require $\rho''(\tilde{T})$ bounded.

Acknowledgments. I am indebted to Professor S. Rionero for helpful discussions on the material presented here. I also very grateful to the organisers of the meeting,

Professors S.N. Antonsev, J.I. Diaz, and S.I. Shmarev for the invitation to lecture at Oviedo, and particularly to Ildefonso Diaz for his kindness.

References

1. Capone, F. & Gentile, M. Nonlinear stability analysis of convection for fluids with exponentially temperature - dependent viscosity. *Acta Mech.*, in the press.

2. Depassier, M.C. & Spiegel, E.A. Convection with heat flux prescribed on the boundaries of the system. I. The effect of temperature dependence of material properties. *Geophys. Astrophys. Fluid Dyn.*, **21** (1982), 167–188.

3. Franchi, F. & Straughan, B. Stability and nonexistence results in the generalized theory of a fluid of second grade. *J. Math. Anal. Appl.*, **180** (1993), 122–137.

4. Gertsberg, V.L. & Sivashinsky, G.I. Large cells in nonlinear Rayleigh - Bénard convection. *Prog. Theor. Phys.*, **66** (1981), 1219–1229.

5. Lide, D.R. *Handbook of Chemistry and Physics.* 72nd ed. C.R.C. Press, Inc., Boca Raton, Florida (1991-92).

6. Oliver, D.S. & Booker, J.R. Planform of convection with strongly temperature dependent viscosity. *Geophys. Astrophys. Fluid Dyn.*, **27** (1983), 73–85.

7. Palm, E., Ellingsen, T. & Gjevik, B. On the occurrence of cellular motion in Bénard convection. *J. Fluid Mech.* **30**, (1967) 651–661..

8. Richardson, L. & Straughan, B. A nonlinear stability analysis for convection with temperature-dependent viscosity. *Acta Mech.*, **97** (1993), 41–49.

9. Richardson, L. & Straughan, B. Convection with temperature dependent viscosity in a porous medium: nonlinear stability and the Brinkman effect. *Atti Accad. Naz. Lincei*, (Ser. IX), **4** (1993), 223–230.

10. Selak, R. & Lebon, G. Bènard - Marangoni thermoconvective instability in presence of a temperature - dependent viscosity. *J. Phys. II France*, **3** (1993), 1185–1199.

11. Stiles, P.J. Electro-thermal convection in dielectric liquids. *Chemical Phys. Letters*, **179** (1991), 311–315.

12. Straughan, B. Stability criteria for convection with large viscosity variations. *Acta Mech.*, **61** (1986), 59–72.

13. Straughan, B. *The energy method, stability, and nonlinear convection.* Vol. **91**, Appl. Math. Sci. Ser., Springer-Verlag (1992).

14. Straughan, B. *Mathematical aspects of penetrative convection.* Pitman Res. Notes Math., **288** (1993), Longman, Harlow.

15. Straughan, B. Energy stability analyses for convection in fluids with temperature dependent viscosity. To appear in *Proc. VIIth International Conf. on Waves and Stability in Continuous Media*, Bologna, 1993.

16. Tippelskirch, H. Über Konvektionszellen insbesondere im flüssigen Schwefel. *Beiträge zur Physik der Atmosphare*, **29**, (1956), 37–54.

17. Torrance, K.E. & Turcotte, D.L. Thermal convection with large viscosity variations. *J. Fluid Mech.* **47** (1971), 113–125.

18. White, D.B. The planforms and onset of convection with a temperature - dependent viscosity. *J. Fluid Mech.* **191** (1988), 247–286.

ENERGY METHODS IN MAGNETOHYDRODYNAMICS

H. TASSO
Max-Planck-Institut für Plasmaphysik, Euratom Association
85748 Garching bei München, Germany

Abstract. A brief summary of energy methods for linear stability in dissipative magnetohydrodynamics is given. In this case, the methods are equally efficient for fixed and free boundary problems. Linear asymptotic stability has implications in nonlinear stability, at least for a modest but finite level of perturbations.

Sufficient conditions for nonlinear stability of dissipative magnetohydrodynamics flows are obtained and applied to the time dependent magnetized Couette flow. The fluid has a plate as boundary and nonlinear stability is unconditional. The range of stable Reynolds numbers is rather modest i.e. of the order of $2\pi^2 \approx 20$.

Nonlinear stability of force free fields can be treated very successfully for all values of dissipation and all levels of perturbations. It requires, however, the presence of perfectly conducting fixed boundaries. Finally, a special inertia-caused Hopf bifurcation is identified and illustrated by an appropriate example.

1. Introduction

Energy or more generally Lyapunov's methods are an elegant tool in the study of stability of fluids [1] . Though their efficiency in the stability of shear flows was rather limited, the applications in the area of magnetohydrodynamics (MHD) were not only clear and elegant, but led to important tools like the 'Energy Principle' of ideal MHD [2] and subsequent work for dissipative MHD [3] . This is certainly due to the nontrivial stability problems of MHD equilibria without or with small flow.

Energy methods for fluids with free boundaries need a special attention, however, in Hydrodynamics (HD) as well as in MHD. The 'Energy Prin-

S. N. Antontsev et al. (eds.), Energy Methods in Continuum Mechanics, 158–174.

ciple' of MHD [2] is derived for a free plasma vacuum interface, and the dissipative 'Energy Principle' [3] can be made so by making the resistivity infinite outside the plasma core. This happy situation is valid only for the linearized stability problem.

Energy methods for nonlinear stability of free boundary problems seem to be very hard to establish. None is known to me in MHD . Even worse is the situation of compressible MHD. Compressibility causes large difficulties in terms of nonvanishing surface integrals even for the fixed boundary problem. This paper is organized as follows : Section 2 summarizes briefly the status of energy methods in linearized resistive MHD . Nonlinear stability of dissipative MHD flows with fixed boundaries is the subject of section 3 . Section 4 specializes on the important case of MHD equilibria without flows . Hopf bifurcations due to overstability are treated in section 5 . Finally, conclusions and outlook are given in section 6 .

2. Energy methods in linearized resistive MHD

First, the equations for MHD equilibrium and the linearized resistive MHD dynamics around equilibrium are given. Then, a sufficient stability condition with respect to purely growing modes is derived (see [4]). Limiting cases of physical interest and implications for nonlinear stability are also discussed.

Resistive MHD equilibria generally have a flow which, for simplicity, we neglect in the equation of motion, but which we keep in Ohm's law. The equilibrium equations are given by

$$\mathbf{J} \times \mathbf{B} = \nabla P_0, \tag{1}$$

$$\nabla \cdot \mathbf{B} = 0, \tag{2}$$

$$\mathbf{E} + \mathbf{V} \times \mathbf{B} = \eta_0 \mathbf{J}. \tag{3}$$

As usual $\mathbf{B}$ is the magnetic field, $\mathbf{J} = \nabla \times \mathbf{B}$, $\mathbf{E}$ is the curl-free electric field, $\mathbf{V}$ is the flow velocity due to resistivity η_0 and P_0 is the pressure. The 'existence' of magnetic surfaces is assumed and the resistivity is taken as constant on these surfaces. The equations of the linearized perturbations are

$$\rho\ddot{\xi} + \nabla P_1 - \mathbf{j} \times \mathbf{B} - \mathbf{J} \times \mathbf{b} = 0, \tag{4}$$

$$\mathbf{e} + \dot{\xi} \times \mathbf{B} + \mathbf{V} \times \mathbf{b} - \eta_1 \mathbf{J} - \eta_0 \mathbf{j} = 0, \tag{5}$$

$$\nabla \times \mathbf{e} = -\dot{\mathbf{b}}, \tag{6}$$

$$\nabla \cdot \mathbf{b} = 0, \tag{7}$$

$$\mathbf{j} = \nabla \times \mathbf{b}, \tag{8}$$

$$\mathbf{B} \cdot \nabla \eta_1 + \mathbf{b} \cdot \nabla \eta_0 = 0, \tag{9}$$

$$P_1 = -\gamma P_0 \nabla \cdot \xi - \xi \cdot \nabla P_0, \qquad (10)$$

where ρ is the mass density, P_1, $\mathbf{j}$, $\mathbf{b}$, $\mathbf{e}$ and η_1 are the perturbations of, respectively, pressure, current, magnetic field, electric field and resistivity. The boundary conditions are $\mathbf{n} \cdot \mathbf{b} = \mathbf{n} \cdot \xi = \mathbf{0}$, where $\mathbf{n}$ is the normal to a perfectly conducting wall.

Let us express $\mathbf{e}$ and $\mathbf{b}$ in terms of the vector potential $\mathbf{A}$ and take the gauge of zero scalar potential :

$$\mathbf{e} = -\dot{\mathbf{A}},$$
$$\mathbf{b} = \nabla \times \mathbf{A},$$

with the boundary condition $\mathbf{n} \times \mathbf{A} = \mathbf{0}$. We insert $\mathbf{j}$ from eq.(5) into eq.(4) to obtain a system written in terms of $\Psi = \begin{pmatrix} \xi \\ \mathbf{A} \end{pmatrix}$:

$$N\ddot{\Psi} + P\dot{\Psi} + Q\Psi = 0, \qquad (11)$$

where N, P and Q are given by, respectively,

$$N = \begin{pmatrix} \rho & 0 \\ 0 & 0 \end{pmatrix},$$

$$P = \begin{pmatrix} \mathbf{B}/\eta_0 \times (\cdots \times \mathbf{B}) & (\cdots \times \mathbf{B}/\eta_0) \\ -(\cdots \times \mathbf{B}/\eta_0 & 1/\eta_0 \end{pmatrix},$$

and

$$Q = \begin{pmatrix} \begin{array}{c} \nabla(-\gamma P_0(\nabla \cdot \ldots)) \\ -\nabla(\ldots \cdot \nabla P_0) \end{array} & \begin{array}{c} -\mathbf{J} \times (\nabla \times \cdots) \\ -1/\eta_0 \nabla P_0 (\mathbf{B} \cdot \nabla)^{-1} (\nabla \times \ldots \cdot \nabla \eta_0) \\ +\mathbf{B}/\eta_0 \times (\mathbf{V} \times \nabla \times \cdots) \end{array} \\ & \\ 0 & \begin{array}{c} \nabla \times \nabla \cdots \\ +\mathbf{J}/\eta_0 (\mathbf{B} \cdot \nabla)^{-1} (\nabla \times \ldots \cdot \nabla \eta_0) \\ -\mathbf{V}/\eta_0 \times \nabla \times \cdots \end{array} \end{pmatrix}.$$

The first two matrix operators are symmetric and positive. The last operator Q is obviously not selfadjoint. For this reason we cannot find a Lyapunov functional which would lead to a necessary and sufficient condition for stability as in, for example, [5] or [6].

As shown in [4], one can, however, write a sufficient condition for stability against purely growing modes in the form

$$\delta W = (\Psi, Q\Psi) \geq 0, \qquad (12)$$

where the scalar product is defined with purely real quantities. Only the symmetric part Q_S of Q survives in eq.(12), but if a symmetrized form for eq.(12) is wanted, it is easy to construct Q^+, the adjoint of Q, by integration by parts, and use $Q_S = (Q + Q^+)/2$ instead of Q in eq.(12).

Criterion (12) implies volume integrations which can be reduced to integrations on the magnetic surfaces and integrations across them. The operator $(\mathbf{B} \cdot \nabla)^{-1}$ in eq.(12), which comes from integration of eq.(9), is singular across the rational surfaces ($1/x$ singularity). This singularity is physically prohibited by the breakdown of eq.(9) due to a finite heat conduction κ ($\kappa_{\|}$ is assumed to be infinite and $\kappa_{\perp} = 0$ for eq.(9)). In fact, η_1 should not become infinite on the rational magnetic surfaces, but small. It is then natural to define the integrations across the surfaces in the sense of Cauchy principal parts (no delta functions) as in [6]. Note here that these singularities are not aggravated by the above-mentioned symmetrizing integrations by parts, because they occur on the surfaces.

Let us now write δW explicitly:

$$\begin{aligned} \delta W \;=\; & \int d\tau(\gamma P_0(\nabla \cdot \xi)^2 + (\xi \cdot \nabla P_0)\nabla \cdot \xi) \\ & + \int d\tau(\nabla \times \mathbf{A})^2 - \int d\tau \xi \times \mathbf{J} \cdot \nabla \times \mathbf{A} + \\ & + p\int d\tau \mathbf{J} \cdot (\mathbf{A} - \xi \times \mathbf{B})(\mathbf{B} \cdot \nabla)^{-1}(1/\eta_0)(\nabla \eta_0 \cdot \nabla \times \mathbf{A}) \\ & - \int d\tau(\mathbf{A} - \xi \times \mathbf{B}) \cdot \mathbf{V} \times (\nabla \times \mathbf{A})1/\eta_0. \end{aligned} \quad (13)$$

If we choose in δW the MHD test function $\mathbf{A} = \xi \times \mathbf{B}$, then δW reduces to δW_{MHD}(see [2]). In the tokamak scaling (large axial wavelength and magnetic fields) and for $\mathbf{J} = \mathbf{e_z} J$, $\eta_0 J = ct\cdot$, $\xi = \mathbf{e_z} \times \nabla U$, $\mathbf{V} = 0$, δW reduces to the necessary and sufficient condition found in [6] for fixed boundary. If, in addition, we make the resistivity infinite outside a certain plasma core, we obtain the equivalent of a vacuum in that region, and a full 'Energy Principle' with free boundary is obtained.

It is more convenient to treat δW in Hamada-like coordinates especially for the term $(\mathbf{B} \cdot \nabla)^{-1}$, which also appears in [6]. The symmetrization of Q, if desired, can be done either analytically in the same coordinates by integration by parts or after discretization in the case of numerical evaluation by computing the adjoint matrix.

The equilibrium quantities in eq.(13) should satisfy equations (1)-(3). To determine the contribution of the last integral in eq.(13), one requires a knowledge of unavoidable [7] Pfirsch-Schlüter- like flows, which are important especially for stellarators. The flow in a tokamak can probably be neglected if the aspect ratio is large enough and the poloidal currents are weak. One can then take $\nabla \times \eta_0 \mathbf{J} \approx \mathbf{0}$ as in [6].

The main advantage of (13) is that it can be numerically evaluated by spectral methods well known in ideal MHD stability and recently extended to MHD stability of stellarator equilibria. A second positive aspect is that this approach to resistive MHD stability is the only one which takes real geometry into account together with the complex flows it generates, and in an exact way at that.

One can 'upgrade' conditions (12)-(13) for two limiting cases of physical interest : 1) for $Q_a \approx \epsilon$ small, which relates to the 'tokamak scaling'. The condition becomes necessary and sufficient for stability with respect to all modes (not the purely growing only). 2) $N = 0$, or neglecting inertia in which case the conditions (12)-(13) become sufficient for stability with respect to all modes. In addition, simplified versions of (13) can be obtained through physical or natural choices of the test function space. These points are discussed in detail in [3] and [19].

Before we proceed to the full nonlinear stability problem, let us mention that linear asymptotic stability of dissipative systems imply nonlinear stability for a restricted level of perturbations. In other words, linear stability of such systems provides already some finite basin of attraction in functional space. The boundaries delivered by the estimates theorems may be, however, much smaller than actual boundaries as will be seen in the next sections. For a discussion of such estimates for fluids see [8].

3. Nonlinear MHD stability with flow

Following the papers [1] , [9] , [10] , [11] , it is possible to formulate a sufficient condition for nonlinear stability in HD and MHD . It turns out that, in case it is satisfied, the system is unconditionally stable i.e. for all levels of perturbations in functional space. The condition is, however, satisfied for rather small Reynolds numbers of the order of 20. Let us give here a derivation, which, though not rigorous in the pure mathematical sense, is basic and compact. For incompressible fluids and in particular in HD and MHD the nonlinear terms in the equation of motion are of the quasilinear type and dissipation is present in the form of material viscosity or resistivity. More precisely if $\underline{u}$ is a many components vector field in an L^2 function space representing the frame of the fluid motion, $\underline{u}$ will obey an equation of the form

$$\dot{\underline{u}} = A(\underline{u})\underline{u} + D\underline{u}, \tag{14}$$

where $A(\underline{u})$ is a nonlinear operator depending linearly upon $\underline{u}$ and D is a linear negative definite operator if $\underline{u} = 0$ at the boundary. A simple example is

$$A(\underline{u})\underline{u} = \underline{u} \cdot \nabla \underline{u}\,, \qquad D\underline{u} = \nabla^2 \underline{u}. \tag{15}$$

We assume further that

$$(\underline{u}, A(\underline{u})\underline{u}) = 0, \tag{16}$$

where the scalar product is given by

$$(\underline{a}, \underline{b}) = \int \underline{a} \cdot \underline{b}\, d\tau, \tag{17}$$

the integration being done over the volume occupied by the fluid.

To study the nonlinear stability we split $\underline{u}$ in

$$\underline{u} = \underline{u}_0 + \underline{u}_1, \tag{18}$$

where $\underline{u}_1$ is a finite perturbation zero at the boundary and $\underline{u}_0$ satisfies

$$\dot{\underline{u}}_0 = A(\underline{u}_0)\underline{u}_0 + D\underline{u}_0. \tag{19}$$

The equation for $\underline{u}_1$ is then

$$\dot{\underline{u}}_1 = A(\underline{u}_1)\underline{u}_1 + L\underline{u}_1, \tag{20}$$

with

$$L\underline{u}_1 = A(\underline{u}_0)\underline{u}_1 + A(\underline{u}_1)\underline{u}_0 + D\underline{u}_1. \tag{21}$$

L is a linear operator on $\underline{u}_1$ which in cases like (15) will remain negative definite if $A(\underline{u}_0)$ and $\underline{u}_0$ are small enough. Taking the scalar product of $\underline{u}_1$ with equation (20) we obtain

$$\frac{1}{2}(\underline{u}_1, \underline{u}_1)^{\cdot} = (\underline{u}_1, L\underline{u}_1) \tag{22}$$

by virtue of (16). Since all considered quantities are real we have

$$(\underline{u}_1, L\underline{u}_1) = (\underline{u}_1, L_s\underline{u}_1)\,, \tag{23}$$

where L_s is the symmetric part of L. Nonlinear stability is then warranted by Lyapunov methods if

$$(\underline{u}_1, L_s\underline{u}_1) < 0, \tag{24}$$

for all $\underline{u}_1$ satisfying $(\underline{u}_1, \underline{u}_1) =$ finite and $\underline{u}_1 = 0$ at the boundary. Expression (24) is a sufficient condition for nonlinear stability. The stability problem is now reduced to the minimization of the hermetian form $(\underline{u}_1, L_s\underline{u}_1)$. This can always be done for any flow ultimately numerically using standards hermiteans eigenvalues techniques.

3.1. APPLICATION TO MHD COUETTE FLOWS

Let us illustrate the above procedure by studying the nonlinear stability of a time-dependent MHD flow generalising the time-dependent planar Couette flow. It consists of a fluid bounded by two horizontal plates, the first plate at $z = 0$ and the second at $z = h$, with velocity parallel to the magnetic field and both depending only on one coordinate (z) and the time (t):

$$\mathbf{v} = v(z,t)\,\hat{e}_y, \tag{25}$$

$$\mathbf{B} = B(z,t)\,\hat{e}_y, \tag{26}$$

satisfying the equations

$$\frac{\partial v}{\partial t} - \nu \frac{\partial^2 v}{\partial z^2} = 0, \tag{27}$$

$$\frac{\partial B}{\partial t} - \eta \frac{\partial^2 B}{\partial z^2} = 0, \tag{28}$$

$$\frac{\partial p}{\partial z} + B \frac{\partial B}{\partial z} = 0, \tag{29}$$

where ν and η are viscosity and resistivity respectively. For simplicity special solutions of these equations can be taken as

$$\mathbf{v} = \frac{v_0}{\sin\sqrt{\frac{\alpha}{\nu}}h} e^{-\alpha t} \sin\sqrt{\frac{\alpha}{\nu}}z\,\hat{e}_y, \tag{30}$$

$$\mathbf{B} = \frac{B_0}{\sin\sqrt{\frac{\alpha}{\eta}}h} e^{-\alpha t} \sin\sqrt{\frac{\alpha}{\eta}}z\,\hat{e}_y, \tag{31}$$

$$p = -\frac{B^2}{2} + f(t), \tag{32}$$

with the following boundary conditions:

$$v(0,t) = B(0,t) = 0, \tag{33}$$

$$v(h,t) = v_0\,e^{-\alpha t}, \tag{34}$$

$$B(h,t) = B_0\,e^{-\alpha t}. \tag{35}$$

and $f(t)$ fixed by the boundary conditions on p. In the limit $\alpha \to 0$ this system reduces to a stationary MHD flow. For $B_0 \to 0$ we have the time dependent Couette flow and when both α and $B_0 \to 0$, we obtain the stationary Couette flow.

Without going into the details of the calculations and notations, which can be found in [11], we give an outline of the successive steps leading to the final result. We obtain first the operator L from (21) and the nonlinear MHD equations, then extract the symmetric part of L, L_s and apply criterion (24) by maximizing the left hand side of (24). The final results can be summarized as follows

– For $\nu < \eta$ the system is stable if

$$Re\frac{\sqrt{\frac{\alpha}{\nu}}\,h}{\sin\sqrt{\frac{\alpha}{\nu}}\,h} + S\frac{\sqrt{\frac{\alpha}{\eta}}\,h}{\sin\sqrt{\frac{\alpha}{\eta}}\,h} < 2\pi^2, \tag{36}$$

where

$$Re = \frac{v_0 h}{\nu}, \tag{37}$$

$$S = \frac{B_0 h}{\nu}. \tag{38}$$

– For $\eta < \nu$ the system is stable if

$$Re_m\frac{\sqrt{\frac{\alpha}{\nu}}\,h}{\sin\sqrt{\frac{\alpha}{\nu}}\,h} + S_m\frac{\sqrt{\frac{\alpha}{\eta}}\,h}{\sin\sqrt{\frac{\alpha}{\eta}}\,h} < 2\pi^2, \tag{39}$$

where

$$Re_m = \frac{v_0 h}{\eta}, \tag{40}$$

$$S_m = \frac{B_0 h}{\eta}. \tag{41}$$

In the limit $\alpha \to 0$ (steady MHD flow) we obtain

$$Re + S < 2\pi^2 \quad for \quad \nu < \eta \tag{42}$$

and

$$Re_m + S_m < 2\pi^2 \quad for \quad \eta < \nu. \tag{43}$$

For the time-dependent Couette flow ($B_0 \to 0$), we have

$$Re\frac{\sqrt{\frac{\alpha}{\nu}}\,h}{\sin\sqrt{\frac{\alpha}{\nu}}\,h} < 2\pi^2, \tag{44}$$

and for the stationary Couette flow ($\alpha, B_0 \to 0$)

$$Re < 2\pi^2. \tag{45}$$

The critical value $2\pi^2 \simeq 19.7$ for the Reynolds number calls for some comments. The extremalization for the Couette flow in HD has been done in the literature by constraining the variations to be divergence-free (see e.g. [12]). As a consequence of that the critical value of 20.7 is found. This gain of 5% in the critical value is paid by a very sophisticated derivation which would not be tractable in the case of the generalized unsteady MHD Couette flow considered here. This justifies our procedure, which allows compressible test functions for the extremalization.

The sufficient condition (24) is general and robust, but also too stringent. It is fulfilled in HD and MHD only if the Reynolds and magnetic Reynolds numbers are small enough. Since viscosity and resistivity especially for hot plasmas are small, condition (24) would allow only a very low level of electrical currents and flows. Linear stability analysis and experimental evidence, however, seem to show that in some cases, values for currents and flows far beyond those allowed by condition (24) occur without any sign of gross instabilities. In the next section force free fields will be shown unconditionally stable for any value of the resistivity.

4. Nonlinear stability of MHD equilibria without flow

The stability of complex systems such as fluids or plasmas is usually investigated in the linearized case. Obviously, the linearization is done in order to simplify the analysis and obtain a first insight into the problem. This is, however, by no means sufficient for practical stability for the following reasons : If a system is linearly stable, it implies stability only for infinitesimal perturbations. If it is linearly unstable, it may saturate at a low or high level in the nonlinear regime. Since for practical situations the perturbations are finite and the saturation levels critical, the study of nonlinear stability, especially for fluids and plasmas, becomes an important and sometimes crucial issue.

In hydrodynamics (HD) the planar Couette flow and the Poiseuille flow in a circular pipe are both linearly stable for all Reynolds numbers (see [12] and [13]). In practical situations turbulence occurs at Reynolds numbers larger than roughly one thousand. It is attributed to nonlinear instabilities or instabilities due to finite perturbations. This view was lent support by simple amplitude expansions [13] and numerical calculations [14] .

In HD and magnetohydrodynamics (MHD) exact sufficient criteria for nonlinear stability exist (see [1], [10] and [11]). Such criteria are powerful and robust, and provide nonlinear stability for arbitrary perturbation levels. In other words, they ensure so-called unconditional stability, which in a certain sense is too good and is not needed for practical stability, since the perturbations can be assumed to be limited in an experiment, especially if one wants to avoid strong vibrations etc. Accordingly, the critical Reynolds numbers delivered by these criteria are too low, of the order of 5 (see [1]) and 20 (see [11]).

Unfortunately, no rigorous criteria are available in HD in the range of Reynolds numbers larger than roughly 20 . This lack of knowledge is precisely in the range where the nonlinear stability margin will probably depend upon the perturbation level. This is equivalent to saying that what is missing is a knowledge of the basin of attraction of the unperturbed solution in functional space. For very low Reynolds numbers 5 to 20 the basin of attraction is infinite and for very large Reynolds numbers it is probably infinitesimal or very small.

Fortunately, the situation is not as bad in MHD. It is possible there to find unperturbed equilibria with zero flow which are unconditionally stable for all magnetic Reynolds numbers. A first example is the case of so-called force-free fields, whose nonlinear stability was recently analyzed by the author [15].

4.1. NONLINEAR STABILITY OF FORCE-FREE FIELDS

Let us assume as unperturbed solution

$$\mathbf{j} = \lambda \mathbf{B} \tag{46}$$

with $\lambda = ct.$ bounded by a perfectly conducting wall.

The equations of motion are those of incompressible MHD with a material resistivity η constant in space and time. For any finite perturbations $\mathbf{v}$ of the velocity field with $\mathbf{n} \cdot \mathbf{v} = 0$ at the boundary and $\mathbf{A}$ of the vector potential with $\mathbf{n} \times \mathbf{A} = 0$ at the boundary the equations of motion are [15]

$$\dot{\mathbf{v}} + \mathbf{v} \cdot \nabla \mathbf{v} = \mathbf{J}_0 \times \mathbf{B}_1 + \mathbf{j}_1 \times \mathbf{B}_0 + \mathbf{j}_1 \times \mathbf{B}_1, \tag{47}$$

with $\nabla \cdot \mathbf{v} = 0$,

$$\dot{\mathbf{A}} = \mathbf{v} \times (\mathbf{B}_0 + \mathbf{B}_1) - \eta \mathbf{j}_1, \tag{48}$$

$$\dot{\mathbf{B}}_1 = \nabla \times (\mathbf{v} \times (\mathbf{B}_0 + \mathbf{B}_1) - \eta \mathbf{j}_1). \tag{49}$$

Taking the scalar product of (47) with $\mathbf{v}$ and that of (49) with $\mathbf{B}_1$, adding and integrating over the volume, we obtain

$$\frac{\partial}{\partial t} \int \frac{d\tau}{2} (v^2 + \mathbf{B}_1^2) = \lambda \int d\tau \mathbf{v} \times \mathbf{B}_0 \cdot \mathbf{B}_1 - \int d\tau \eta \mathbf{j}_1^2. \tag{50}$$

Many quadratic and cubic terms integrate to zero because of the boundary condition being taken as perfectly conducting. Taking the scalar product of (48) with $\mathbf{B}_1$, we can solve for $\mathbf{v} \times \mathbf{B}_0 \cdot \mathbf{B}_1$, and, inserting into (50), we obtain

$$\frac{\partial}{\partial t}\int \frac{d\tau}{2}(v^2 + \mathbf{B}_1^2 - \lambda \mathbf{A} \cdot \nabla \times \mathbf{A}) = -\eta \int d\tau (\mathbf{j}_1^2 - \lambda \mathbf{B}_1 \cdot \mathbf{j}_1) \quad (51)$$

or

$$\frac{d}{dt}\int d\tau \frac{1}{2}(\mathbf{v}^2 + (\nabla \times \mathbf{A})^2 - \lambda \mathbf{A} \cdot \nabla \times \mathbf{A}) = -\eta \int d\tau ((\nabla \times \nabla \times \mathbf{A})^2 - \lambda \nabla \times \mathbf{A} \cdot \nabla \times \nabla \times \mathbf{A}). \quad (52)$$

Since (52) also holds for the linearized case, which was discussed a long time ago in [16], we reproduce the proof given there for the sufficiency of

$$\int d\tau \frac{1}{2}((\nabla \times \mathbf{A})^2 - \lambda \mathbf{A} \cdot \nabla \times \mathbf{A}) \geq 0 \quad (53)$$

for nonlinear stability. Note first that $\mathbf{n} \times \mathbf{A} = 0$ implies $\mathbf{n} \cdot \nabla \times \mathbf{A} = 0$ at the boundary, so that if (53) is satisfied for $\mathbf{n} \times \mathbf{A} = 0$, then the right-hand side of (52) will be satisfied for $\mathbf{n} \cdot \nabla \times \mathbf{A} = 0$. By means of the Lyapunov theorems the expression under the time derivative of the left-hand side of (52) is a Lyapunov function if (53) is verified. Condition (53) is sufficient for stability independently of the values of the resistivity and viscosity. As mentioned above, there is nothing like this in HD .

4.2. REMARK

Though the derivation of the stability condition in Reference [16] is basically correct, some technical imperfections can be removed. The minimization of the square brackets appearing on the second member of equation (26) of [16] should have been done by restricting $\mathbf{B}_1$ and its variations to be curl of some vector potential $\mathbf{A}$ (see Acknowledgements).

This causes a $\nabla\Phi$ term to appear in equation (29) of [16]. This term, however, can be proved to vanish if $\mathbf{B}_1$ vanishes at the boundary, which is a kind of "hydrodynamical" boundary condition, more stringent than the condition at a perfect conductor.

4.3. TWO-DIMENSIONAL PERTURBATIONS

A less spectacular example in MHD is the nonlinear stability of a straight z-pinch or tokamak surrounded by perfectly conducting walls. Here it is possible to prove nonlinear stability with respect to 2-dimensional perturbations if the current density is homogeneous, the velocity of the unperturbed fluid $\mathbf{v}_0$ being zero. The equilibrium is given by

$$\Delta\Psi = J_0 = -P'(\Psi), \tag{54}$$
$$\mathbf{j}_0 = \mathbf{e}_z J_0, \tag{55}$$
$$\mathbf{v}_0 = 0. \tag{56}$$

Ψ denotes the flux of the poloidal magnetic field, J_0 is the current density in the z direction and $P(\Psi)$ is the pressure as a function of Ψ. A constant magnetic field B_z in the z direction could be added without changing the shape of Ψ, which is determined by (54) for any given boundary condition on Ψ.

The MHD equations of motion for an incompressible fluid with mass density equal to unity are

$$\frac{\partial \mathbf{v}}{\partial t} + \mathbf{v}\cdot\nabla\mathbf{v} = \mathbf{j}_1 \times \mathbf{B}_0 + \mathbf{j}_0 \times \mathbf{B}_1 + \mathbf{j}_1 \times \mathbf{B}_1 - \nabla P_1 + \mu\Delta\mathbf{v}, \tag{57}$$
$$-\frac{\partial \mathbf{B}_1}{\partial t} = -\nabla \times (\mathbf{v} \times (\mathbf{B}_0 + \mathbf{B}_1)) + \eta\nabla \times \mathbf{j}_1, \tag{58}$$

where $\mathbf{v}$ and $\mathbf{B}_1$ are finite perturbations of the velocity and the magnetic fields having $\mathbf{n}\cdot\mathbf{v} = \mathbf{n}\cdot\mathbf{B}_1 = 0$ at the boundary. Taking the scalar product of (57) with $\mathbf{v}$ and that of (58) with $\mathbf{B}_1$, adding and integrating over the volume, we obtain

$$\frac{d}{dt}\frac{1}{2}\int(\mathbf{v}^2+\mathbf{B}_1^2)d\tau = \int \mathbf{v}\cdot\mathbf{j}_0\times\mathbf{B}_1 d\tau - \mu\int(\nabla\times\mathbf{v}^2)d\tau - \eta\int(\nabla\times\mathbf{B}_1^2). \tag{59}$$

Many quadratic and cubic terms integrate to zero because of the boundary conditions. The right-hand side of (59) would be negative if the first integral on the right-hand side of (59) vanished. We now prove that this is the case if $\mathbf{j}_0 = ct$. Introducing the vector potential, we have

$$\int \mathbf{v}\times\mathbf{j}_0\cdot\nabla\times\mathbf{A}d\tau = \int \mathbf{A}\cdot(\mathbf{j}_0\cdot\nabla\mathbf{v} - \mathbf{v}\cdot\nabla\mathbf{j}_0)d\tau = 0. \tag{60}$$

The last equality is due to the two-dimensionality of $\mathbf{v}$ and the assumption of a constant vector $\mathbf{j}_0$. This means that the expression under the time derivative on the left-hand side of (59) is a Lyapunov function, from which nonlinear stability follows. Again this condition is independent of the Reynolds and magnetic Reynolds numbers, in contrast to, for example, [11], but the stability is unconditional.

For the above two examples we were able to obtain sufficient and unconditional stability conditions without limitations on the Reynolds numbers. Our examples are of course special and contain an important ingredient $\mathbf{v}_0 = 0$, i.e. no flow in the unperturbed state. This is nontrivial only in the MHD cases. In contrast, the criteria [1], [10] and [11] are very general but imply severe limitations on the Reynolds numbers. The examples given in this note and the general criteria [1], [10] and [11] have one thing in common, viz-they all deal with unconditional stability.

As mentioned at the beginning, this is not necessary for practical stability. If we want to get rid of unconditional stability in our proofs, we have to deal with finite basins of attraction in functional spaces. Practical stability is tied to this very difficult problem.

5. A manifest Hopf bifurcation

In this section we consider the case for which condition (12) is satisfied and prove that, if the inertial term can cause some additional overstability, the modes appearing in this way meet the requirements of the centre manifold theorem [17]. This means that they can be stabilized nonlinearly through a Hopf bifurcation, resulting in a limit cycle or nonlinear periodic oscillation. If $N \neq 0$ in equation (11), inertia-caused overstable modes can occur : In a special example [18] , the overstability occurs only in the compressible case, primarily at the magnetoacoustic resonance.

Let us now consider the case for which (12) is satisfied but (11) is overstable for $N \neq 0$. Any overstable mode of (11) is given by

$$\xi = \Psi e^{(i\omega+\gamma)t}, \tag{61}$$

where ω and γ are real and satisfy

$$(\gamma^2 - \omega^2)(\Psi, N\Psi) + \gamma(\Psi, P\Psi) + (\Psi, Q_s\Psi) = 0, \tag{62}$$

$$2\gamma\omega(\Psi, N\Psi) + \omega(\Psi, P\Psi) + (\Psi, Q_a\Psi) = 0. \tag{63}$$

We see from (62) and (63) and, generally, from the reality of the operators in (11) that $\xi^\star = \Psi^\star e^{(-i\omega+\gamma)t}$ is also an eigenmode of (11). It follows that the modes due to the inertia operator N always come in pairs with opposite sign of the real frequencies but the same growth rate, all other modes being damped because of (12). These features are precisely the principal ingredients of the centre manifold theorem [17]. In summary, if (12) is satisfied, inertia-caused overstability can lead to a Hopf bifurcation resulting in a periodic nonlinear oscillation.

Let us illustrate the occurence of the overstability and the Hopf bifurcation by choosing the operators N and P of (11) proportional to the identity.

$$nI\ddot{\Psi} + pI\dot{\Psi} + (Q_s + Q_a)\Psi = 0, \tag{64}$$

where n and p are positive numbers.

If the eigenvalue problem for $Q_s + Q_a$ can be solved, we have

$$(Q_s + Q_a)\Psi_m = \lambda_m \Psi_m \tag{65}$$

with

$$\lambda_m = \lambda_{mR} + i\lambda_{mI}. \tag{66}$$

λ_m and Ψ_m are, in general, complex, though (64) involves only real quantities.

If we make the ansatz $\Psi = \Psi_m e^{\omega_m t}$, the eigenvalues of (64) ω_m are related to the λ_m by the following equation

$$n\omega^2 + p\omega + \lambda_R + i\lambda_I = 0, \tag{67}$$

valid for each pair of eigenvalues λ_m and ω_m.

Let us split ω in real and imaginary parts, then (67) can be written as a system

$$\begin{aligned} n(\omega_R^2 - \omega_I^2) + p\omega_R + \lambda_R &= o, & (68)\\ \omega_I(2n\omega_R + p) + \lambda_I &= 0. & (69) \end{aligned}$$

Inserting in (68) the value of ω_I obtained from (69), we have

$$n\omega_R^2 + p\omega_R + \lambda_R = \frac{n\lambda_I^2}{(2n\omega_R + p)^2}. \tag{70}$$

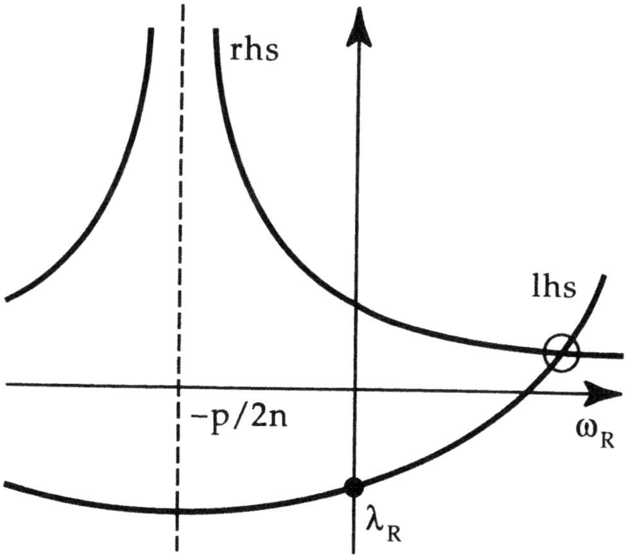

Figure 1 $\lambda_R < 0$

Since n and p are positive, it is easy to make schematic plots of the left hand side (lhs) and right hand side (rhs) of (70) (see Figures 1 and 2). Let us consider two cases : first, some $\lambda_R < 0$ (Fig. 1), and second, all $\lambda_R > 0$ (Fig. 2), which follows from the validity of (12).

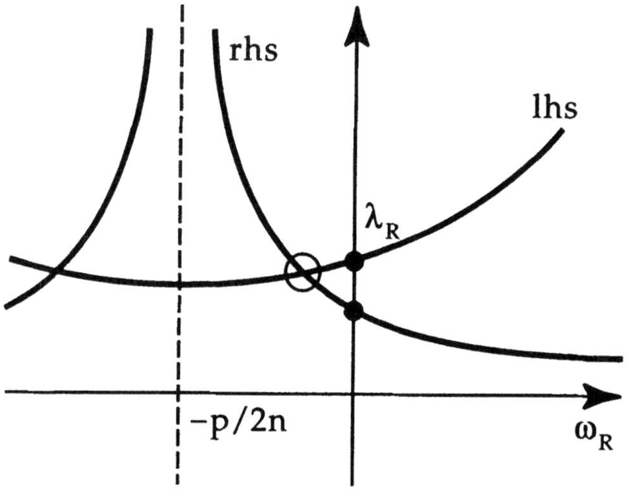

Figure 2 $\lambda_R < 0$

We see that, if some $\lambda_R < 0$, the system (68, 69) has always a positive root ω_R (see Fig. 1), which means instability. Let us note that a violation of (12) for a Ψ which is not representative of the eigenfunction, does not necessarily imply that $\lambda_R < 0$.

In case $\lambda_R > 0$, the crossing point leads to an instability only if $\lambda_R < \frac{n\lambda_I^2}{p^2}$ (see Fig. 2). As mentioned above and in [20] the unstable crossing cannot occur for $n \approx 0$ or λ_I very small or p large. Starting from a small n, for $\lambda_R > 0$ and λ_I and p fixed, we obtain an overstability by increasing the value of n until

$$\lambda_R < \frac{n\lambda_I^2}{p^2}, \tag{71}$$

as explained in [20].

6. Conclusions

Energy methods appear as a powerful tool in MHD stability in the linear case as well as in the nonlinear one. The energy methods of the linear case are efficient for fixed and free boundaries. The linear stability problem is essentially reducible to a Hermitean form, which can be minimized analytically or numerically. For general geometries, codes similar to those considered in [21] lead to many applications.

At present, energy methods are also efficient for nonlinear stability as long as no free boundaries are considered. The ability to find statements, especially for force free fields in general geometry, is possible if the fluid is bounded by a perfectly conducting wall. The same is true for the nonlinear stability of flows in HD and MHD. The assumption of incompressibility seems also essential for the derivation of criteria in nonlinear stability. This situation is, hopefully, only due to technical reasons, perhaps because the known criteria are unconditional. Unconditional stability seems to be theoretically more accessible, though it is not needed experimentally.

7. Acknowledgements

The author is indebted to D. Pfirsch for his critical comments concerning the derivation of Euler equation (29) of [16].

References

1. J. Serrin. *Handbuch der Physik.*Springer Verlag, Berlin, 1959 vol. VIII/1 p. 125.
2. I.B. Bernstein, E.A. Frieman, M.D. Kruskral, R.M. Kulsrud. An energy principle for hydromagnetic stability problems. *Proc. Roy. Soc* A 244 (1958), 17-40.
3. H. Tasso. Linear and nonlinear stability in resistive magnetohydrodynamics. *Annals of Physics* 234 (1994), 211-224.
4. H. Tasso. A sufficient condition in resistive MHD. *Physics Letters* A 147 (1990), 28-30.
5. H. Tasso. Energy principle for two-dimensional resistive instabilities. *Plasma Physics* 17 (1975), 1131-1134.
6. H. Tasso, J.T. Virtamo. Energy principle for 3-d resistive instabilities in shaped cross-section tokamaks. *Plasma Physics* 22 (1980), 1003-1013.
7. H. Tasso. *Lectures on Plasma Physics.*IFUSP/P-181, Sao Paulo, 1979.
8. J. Neustupa. *Workshop on 'Recent Progress in the Theory of the Navier-Stokes Equations and Related Topics'.*Bayreuth, July 1992.
9. S. Rionero. *Ann. Math. Pura Appl.* LXXVI, 75 (1967), LXXVIII, 339 (1968).
10. G.P. Galdi, S. Rionero. *'Weighted Energy Methods in Fluid Dynamics and Elasticity'.* Lecture Notes in Math., Springer Verlag, 1985.
11. H. Tasso, S.J. Camargo. On the nonlinear stability of dissipative fluids. *Nuovo Cimento* B 107 (1992), 733-740.
12. P.G. Drazin, W.H. Reid. *Hydrodynamic Stability.* Cambridge University Press, 1981.

13. J.T. Stuart. in *Transition and Turbulence.* Editor R.E. Meyer, Academic Press, 1981.
14. S.A. Orszag, A.T. Patera. in *Transition and Turbulence.* Editor R.E. Meyer, Academic Press, 1981.
15. H. Tasso. Simplified versions of a stability condition in resistive MHD. *Physics Letters* A 169 (1992), 396-398.
16. H. Tasso. Lecture at *Theoretical and Computational Plasma Physics.* Trieste, p. 321, IAEA, Vienna, 1978.
17. J.E. Marsden, M. Mc Cracken. *The Hopf Bifurcation and its Applications.* Springer Verlag, New York, 1976.
18. H. Tasso, M. Cotsaftis. Instabilite liee a la compressibilite d'un plasma resistif a courants superficiels. *Plasma Physics (J. Nuclear Energy part C)* 7 (1965), 29-36.
19. H. Tasso. On a general stability condition in resistive MHD. *Physics Letters* 161 (1991), 289-291.
20. H. Tasso. A manifest Hopf bifurcation in resistive magnetohydrodynamics. *Physics Letters* 180 (1993), 257-258.
21. W. Kerner, H. Tasso. Stability of multihelical tearing modes in shaped tokamaks. *Physical Review Letters* 49 (1982), 654-657.

GPSR Compliance

The European Union's (EU) General Product Safety Regulation (GPSR) is a set of rules that requires consumer products to be safe and our obligations to ensure this.

If you have any concerns about our products, you can contact us on ProductSafety@springernature.com

In case Publisher is established outside the EU, the EU authorized representative is:

Springer Nature Customer Service Center GmbH
Europaplatz 3
69115 Heidelberg, Germany

Batch number: 10755292

Printed by Printforce, the Netherlands